# World Checklist of
# Dioscoreales
## Yams and their allies

# World Checklist of

# Dioscoreales
## Yams and their allies

**Rafaël Govaerts, Paul Wilkin and Richard M. K. Saunders**
with
Lauren Raz, Oswaldo Téllez Valdés, Hiltje Maas-van de Kamer,
Paul Maas-van de Kamer and Dian-Xiang Zhang

Kew Publishing
Royal Botanic Gardens, Kew

PLANTS PEOPLE
POSSIBILITIES

First published in 2007 by
Royal Botanic Gardens, Kew
Richmond, Surrey, TW9 3AB, UK
www.kew.org

ISBN 978-1-84246-200-3

British Library Cataloguing in Publication Data
A catalogue record for this book is available from the British Library


Typesetting and page layout: Christine Beard
Design by Media Resources,
Royal Botanic Gardens, Kew


Printed in the United States by Edwards Brothers


For information or to purchase all Kew titles please visit
www.kewbooks.com or email publishing@kew.org


All proceeds go to support Kew's work in saving the world's plants for life

# Contents

Introduction . . . . . . . . . . . . . . . . . . . . . . . . . . . . . . . . . . . . . . . . . . . . . . . . . . . . . . . . vi

How to use the Checklist . . . . . . . . . . . . . . . . . . . . . . . . . . . . . . . . . . . . . . . . . . . x

    Structure . . . . . . . . . . . . . . . . . . . . . . . . . . . . . . . . . . . . . . . . . . . . . . . . . . . . . . . . x

    Names . . . . . . . . . . . . . . . . . . . . . . . . . . . . . . . . . . . . . . . . . . . . . . . . . . . . . . . . . . x

    Acceptance of taxa . . . . . . . . . . . . . . . . . . . . . . . . . . . . . . . . . . . . . . . . . . . . . . x

    Geographical distribution . . . . . . . . . . . . . . . . . . . . . . . . . . . . . . . . . . . . . . . x

    Life-forms . . . . . . . . . . . . . . . . . . . . . . . . . . . . . . . . . . . . . . . . . . . . . . . . . . . . . xi

Abbreviations . . . . . . . . . . . . . . . . . . . . . . . . . . . . . . . . . . . . . . . . . . . . . . . . . . . . . xiii

References . . . . . . . . . . . . . . . . . . . . . . . . . . . . . . . . . . . . . . . . . . . . . . . . . . . . . . . . xv

Acknowledgements . . . . . . . . . . . . . . . . . . . . . . . . . . . . . . . . . . . . . . . . . . . . . . . . xviii

Burmanniaceae . . . . . . . . . . . . . . . . . . . . . . . . . . . . . . . . . . . . . . . . . . . . . . . . . . . . . 1

Dioscoreaceae . . . . . . . . . . . . . . . . . . . . . . . . . . . . . . . . . . . . . . . . . . . . . . . . . . . . . 14

Nartheciaceae . . . . . . . . . . . . . . . . . . . . . . . . . . . . . . . . . . . . . . . . . . . . . . . . . . . . . 63

# Introduction

Dioscoreales are a small but systematically and economically significant order of monocotyledons. The most diverse and important element of Dioscoreales is the yam genus, *Dioscorea* L., with over 600 accepted species names worldwide. Many yam species have edible tubers (Coursey 1967). There are three major cultigens, the winged yam (*D. alata* L.), the guinea yams (*D. cayenensis* Lam./ *D. rotundata* Poir.) and the lesser or Asiatic yam *D. esculenta* (Lour.) Burkill.

The affinities of Dioscoreales, the taxa which comprise it and their relationships to each other were uncertain for many years, but significant progress has been made since 1995. It has been shown to be probably most closely related to Pandanales and to comprise three families, Burmanniaceae, Dioscoreaceae and Nartheciaceae (Caddick *et al.* 2002a, b). Until the 1990's, Burmanniaceae were linked to Orchidaceae and Nartheciaceae to Liliaceae respectively. Dioscoreaceae was thought to belong to a group of plant families with reticulate-veined leaves, which were the most closely-related monocots to the dicots. Dioscoreales are found in temperate and tropical regions worldwide, with the highest diversity being in the seasonally dry tropics of Central South America (Knuth 1924), Mexico (Sosa *et al.* 1987, Tellez & Schubert 1994), the Caribbean (Raz 2007), South Africa (Knuth 1924), Madagascar (Burkill & Perrier 1950, Wilkin *et al.* in prep) and Indochina (Prain & Burkill 1936, 1938, Wilkin & Thapyai 2007) where *Dioscorea* species numbers are highest. The yam allies are either plants of the wet tropics, where the other genera of Dioscoreaceae (Burkill 1951, 1960, Burkill & Perrier 1950) and Burmanniaceae (Maas-van der Kamer 1998) occur, or temperate and montane tropical habitats (Nartheciaceae, Tamura 1998).

The papers by Caddick *et al.* (2002 a, b) showed that Dioscoreaceae is characterised by simultaneous microsporogenesis, a climbing habit, leaves with reticulate secondary veins and petioles with a basal and apical swelling (pulvinus). It comprises four genera, *Dioscorea*, *Tacca* J.R.Forst. & G.Forst., *Trichopus* Gaertn. and *Stenomeris* Planch., which can be differentiated using the characters in Table 1.

All of the dioecious taxa of Dioscoreaceae are now placed in *Dioscorea* (Caddick *et al.* 2002b), including *Tamus* L., *Epipetrum* Phil., *Borderea* Miégev., *Higinbothamia* Uline, *Nanarepenta* Matuda and *Rajania* Uline. Burmanniaceae, the sister family to Dioscoreaceae, are mostly achlorophyllous herbs with reduced parallel-veined leaves, or rarely with chlorophyllous leaves in a basal rosette, and an inferior ovary which develops into 3-winged, usually capsular fruit. Its underground organs are often tuberous. Nartheciaceae are relatively unspecialised monocots with rhizomes, narrow, parallel-veined, mostly basal leaves, flowers usually possessing free tepals and capsular fruits. Nartheciaceae are sister to the rest of Dioscoreales.

Many species of the genus *Dioscorea* are or have been sources of food or medicine. The tubers of at least 50 species are sources of dietary starch in a subsistence or an economic context (Coursey 1967). They vary widely in their ease of collection, palatability and degree of chemical protection from herbivory. Some are only used in times of famine, such as *D.*

Table 1. The morphology and distribution of the genera of Dioscoreaceae.

| | *Dioscorea* | *Tacca* | *Trichopus* | *Stenomeris* |
|---|---|---|---|---|
| Underground Organs | Rhizome and/or tuber | Rhizome and/or tuber | Rhizome | Rhizome with small tubers |
| Stem | Present, usually long and twining to the left or right | Absent | Present, erect and bearing a solitary leaf or long and twining to the right | Present, long, twining to the left |
| Petiole | Not sheathing; basal and apical pulvinii present | Sheathing at base; pulvinii not present | Not sheathing; pulvinus at base only | Not sheathing; basal and apical pulvinii present |
| Inflorescence | Simple or compound raceme or spike, bracts small, foliaceous only | Pseudo-umbel with large, foliaceous and long, filiform bracts | Flowers solitary from among a cluster of chaffy bracts | Compound, cymose |
| Flowers | Usually unisexual; tube/hypanthium present or absent, not urceolate except in 7 species in Madagascar | Perfect, tube/hypanthium short, not urceolate | Perfect, tube/hypanthium short, not urceolate | Perfect, tube/hypanthium expanded, urceolate |
| Stamens | Usually erect and always so at filament base, inserted towards centre of receptacle | Erect, inserted towards centre of receptacle | Erect, inserted towards centre of receptacle | Inserted at hypanthium mouth and reflexed into it |
| Fruit | Dry capsule or rarely baccate, not more than 3 times as long as wide, 2 ovules in each of 3 locules | Dehiscent capsule or indehiscent leathery to fleshy berry, unilocular, multiovulate | Indehiscent, leathery, baccate, 3-locular and 6-ovulate (5 ovules abort in *T. sempervirens* (H. Perr.) Caddick & Wilkin hence 1-seeded) | Dry capsule at least 10 times as long as wide; multiovulate, 3-locular |
| Seed | Usually lenticular or ovoid-lenticular, not ridged, with wing all round margin or restricted to base/apex | Prismatic to reniform, longitudinally ridged, wingless | Ovoid, ruminate, wingless | Flattened-oblanceoloid, longitudinally ridged, winged at apex |
| Distribution | Pantropical to temperate areas as far north as S. Canada, N. Europe and the Russian far East and South to Chile, Western Australia and South Africa (Cape Province) | Pantropical but only one species in Africa and one in South America | Eastern Madagascar, South India, Sri Lanka, Peninsular Malaysia and adjacent Thailand | Northern Borneo and the Philippines |

*dumetorum* (Kunth) Pax in Subsaharan Africa (Wilkin 2001). Its tubers are near the soil surface, and thus easy to collect, but need to have toxic alkaloids removed before they can be eaten safely. Others are protected by being more deeply buried, for example many of the endemic species of Madagascar, or the candle yam in Thailand (*D. brevipetiolata* Prain & Burkill). Spiny roots protect the tuber in some forms of *D. esculenta*. In many parts of the tropics wild yams are a vital food source, especially when staple grains run out. Perhaps the best example of this is Madagascar (Jeannoda *et al.* 2003, 2007). Of the ca. 30 endemic species in Madagascar, at least 24 are edible and 12 or more are eaten regularly in different parts of the country when rice stores have been exhausted. There are introduced, cultivated taxa but their use remains small-scale in gardens and the endemic wild yams are preferred.

The picture is rather different in, for example, West Africa and New Guinea, where there is extensive field cultivation of the *D. cayenensis*/*D. rotundata* complex and *D. alata* and *D. esculenta* respectively as staple starch sources (Coursey 1967). However, yams are also taken from the forest in these areas (e.g. Hladik & Dounias 1996, Dounias 2001). In Africa this is often species such as *D. praehensilis* Benth., *D. abyssinica* Hochst ex Kunth., *D. sagittifolia* De Wild. or *D. baya* De Wild., supposed wild relatives of *D. cayenensis*/*D. rotundata*. People also transplant wild forest yams into their gardens (Hildebrand *et al.* 2002). In New Guinea *D. nummularia* Lam. appears to be a semi-domesticate (Malapa *et al.* 2005). Thus there is a continuum among food yams from taxa where human selection has been limited to those which have been extensively modified by it (e.g. Dumont *et al.* 2006).

In economic terms, medicines have perhaps been an even more important product of *Dioscorea* than food. Many species are rich in steroidal saponins. In particular, the New World species *D. mexicana* Schiedw., *D. composita* Hemsl. and *D. floribunda* M. Martens & Galeotti were used in the 1950s and 60s as the source plants for the contraceptive pill (Coursey 1967, Marks 2001). Other taxa rich in steroidal compounds include *D. deltoidea* Wall. ex Griseb., *D. sylvatica* Eckl. and the genus *Tacca*. As well as synthetic human hormones, yam steroids have also been used to make corticosteroid drugs such as cortisone. The range of medicinal uses of steroids is vast (Sparg *et al.* 2004). However, such compounds are today usually synthesised *de novo* rather than precursors being sourced from plant material.

Preliminary data from Thailand (Wilkin & Thapyai 2007) suggest that 17 of 43 (ca. 40%) Dioscoreaceae species should be given IUCN Red List categories of Near Threatened, Vulnerable, Endangered or Critically Endangered (IUCN 2001). In Madagascar and the Comoros (Wilkin *et al.* 2007a, in prep), this figure is 18/42 (ca. 43%). Thus there is a significant need for targeted species conservation. *D. bako* Wilkin *ined.* (Wilkin *et al.* 2007b) is a good example, a large yam which was not discovered by the many botanists of the 19th and 20th centuries who worked in Madagacar. It is edible, and preferred by people living in the small area of central Western Madagascar to which it is endemic. They are already reporting that it is harder to find than it used to be. Burmanniaceae and Nartheciaceae also contain significant numbers of threatened taxa.

The research on yam conservation, ethnobotany, domestication and sustainable use necessitated by this conservation challenge needs to be underpinned by a stable and universally applied nomenclature. There have been numerous floristic studies in last 30 years, especially of *Dioscorea*, but no global treatment of Dioscoreaceae since Knuth (1924). Thus there is a need for a reliable guide to accepted names. There are still too many gaps in our understanding for a global monograph to be produced, especially in (for example), Burmanniaceae and in *Dioscorea* in Central South America, where there appear to be too many accepted species names. In some cases, the nomenclature has not yet been modified to reflect recently acquired systematic data, for example the transfer the species of *Rajania* to *Dioscorea*. This work is in progress (Raz, in prep.), but because it remains unpublished, *Rajania* and its species are presented as accepted names in this checklist. However, there is enough published information for a useful world checklist to be produced through unifying the regionally-based treatments. We hope in particular that this checklist will help to facilitate closer collaboration between applied yam researchers and systematists.

Systematic research will always result in name changes, and overlooked names will continue to be discovered. We do not pretend that this list is complete or entirely correct. We solicit input and constructive criticism. Please send comments by email to **r.govaerts@kew.org** and **p.wilkin@kew.org** or by regular post to the authors at Herbarium, Royal Botanic Gardens, Kew, Richmond, Surrey TW9 3AB, UK. The checklist is also available on the World Wide Web as a searchable database (**www.kew.org/wcsp/**). The database is live and changes based on any comments made on this or the online version can be incorporated. Of course, as soon as this checklist is printed it will become out of date, but the database which underpins it will constantly be updated. Any changes will appear on the online version at the URL above.

Currently the checklist includes 644 accepted species in Dioscoreaceae in 5 genera, 148 accepted species in Burmanniaceae in 16 genera and 34 accepted species in Nartheciaceae in 5 genera.

# How to use the Checklist

## Structure

The checklist is derived from a database encompassing 24 fields and complying with the data standards proposed by the Taxonomic Databases Working Group (TDWG). Its compilation was effected using Microsoft Access 2002 for Windows, within which editing was also carried out. The arrangement of genera being alphabetical with accepted and synonymous genera intercalated.

## Names

Names of accepted genera and their species and infraspecific taxa are listed alphabetically. Synonymised genera (and species) are intercalated. For each accepted taxon, associated synonyms are listed chronologically if heterotypic, with any homotypic synonyms following in a given lead; in addition, all synonyms in an accepted genus are listed alphabetically at the end of that genus. Place and date of publication of all names are given. Citation of authors follows *Authors of Plant Names* (http://www.ipni.org/ipni/query_author.html); for book abbreviations, the standard is *Taxonomic Literature*, 2nd edn. (Stafleu & Cowan 1976–88; supplements, 1992–2000) also available electronically (http://tl2.idcpublishers.info/); and periodicals are abbreviated according to *Botanico-Periodicum-Huntianum/ 2* (Bridson 2004). A question mark (?) following a name and author indicates that a place of publication has yet to be established. Names of hybrids are preceded by a multiplication sign (?), with the place of publication being followed by the names of the parents if known. Basionyms or replaced synonyms of accepted names are designated by an asterisk (*). For genera, the number of accepted species and the geographical distribution are furnished together with general comments and the suprageneric taxa as used at Kew.

## Acceptance of taxa

Acceptance of species and infraspecific taxa is based not only on assessments of literature but also, where possible, by reference to specialist advice and (where necessary) to the herbarium or living collections. At species level there have historically been in some genera differences of opinion with respect to limits as well as what infraspecific taxa (if any) should be recognised. Initial description of species was often made from plants in cultivation.

## Geographical Distribution

Distributions of **species** and taxa of lower rank are furnished in two ways: firstly by a generalised statement in narrative form, and secondly as TDWG geographical codes (Brummitt, 2001) also available electronically (http://www.nhm.ac.uk/hosted_sites/tdwg/geogrphy.html) expressed to that system's third level. Examples of the former include:

E. & C. U.S.A.
Texas to C. America
Mexico (Veracruz)
Europe to Iran

E. Himalaya, Tibet, China (W. Yunnan)
Philippines (Luzon)
S. Trop. America

When the presence of a taxon in a given region or location is not certainly known, a question mark is used, e.g. New Ireland ?; when an exact location within a country is not known, a question mark within brackets is used, e.g. Mexico (?). Distributions of **genera** are furnished in a relatively simplified form, any special features being given within brackets.

With respect to the TDWG codes, the **region** is indicated by the two-digit number (representative of the first two levels), the first digit also indicating the continent. The letter codes following the digits, when given, represent the third-level **unit** (a country, state or other comparable area). They usually are the first three letters of a given unit's name, but sometimes are contractions. If the country code is not known, '+' is used. For taxa that are known or appear to be extinct in a given region, '†' is used after the country code. Naturalisation is expressed by putting the third-level codes in lower case and, if in a second-level region all occurrences are the result of naturalisation, the code number for the region is placed in brackets. The application of question marks is as indicated above for geographical regions. Examples include:

| | |
|---|---|
| 12 SPA | [SW. Europe: Spain] |
| 32 + | [C. Asia (more exact distribution not known)] |
| 36 CHN? 38 JAP KOR | [Doubtful in China and Eastern Asia; China, Japan, Korea] |
| 51 NZN NZS | [New Zealand: North and South Islands] |
| 76 ARI 77 NWM TEX | [SW. & SC. U.S.A., Mexico & C. America: |
| 79 ALL 80 GUA HON | Arizona, New Mexico, Texas, Mexico, Guatemala and Honduras] |
| 77 TEX† | [SC. U.S.A.: Texas, where extinct] |

## Life-forms

The terminology for *life-forms*, definitions of which follow, is based on the system of Raunkiær (1934, especially chapters 1 and 2) with modifications derived from *Flora van België, het Groothertogdom Luxemburg, Noord-Frankrijk en de aangrenzende gebieden* (De Langhe *et al.* 1983: pp. xvii-xviii, 869 (fig. 16)).
Main Categories:

**phanerophyte** (*phan.*)
  stems: woody and persisting for several years
  buds: normally above 3 m
  e.g. small and large trees

**nanophanerophyte** (*nanophan.*)
  stems: woody and persisting for several years
  buds: above soil level but normally below 3 m
  e.g. shrubs

**chamerophyte** (*cham.*)
  stems: herbaceous and/or woody and persisting for several years
  buds: on or just above soil level, never above 50 cm

**hemicryptophyte** (*hemicr.*)
  stems: herbaceous, often dying back after the growing season, with shoots at soil
    level surviving
  buds: just on or below soil level
  e.g. *Burmannia foliosa*

**geophyte**
  hemicryptophytes that survive unfavourable seasons in the form of a rhizome, bulb, tuber
    or rootbud.

**hydrophyte**
  stems: vegetative shoots sunk in water
  buds: permanently or temporarily on the bottom of the water

**therophyte** (*ther.*)
  plants that survive unfavourable seasons in the form of seeds and complete their life-history
    during the favourable season.
  e.g. annuals


## ADDITIONAL INFORMATION

**climbing** (*cl.*)
  e.g.:
  cl. tuber geophyte: *Dioscorea communis*
  (cl.): scrambling

**parasitic** (*par.*)

**hemiparasitic** (*hemipar.*)
  parasitic plants that are still able to photosynthesise
  e.g. *Viscum album* is a hemipar. nanophan.

**holoparasitic** (*holopar.*)
  parasitic plants that are fully dependent on their host
  e.g. *Orobanche ramosa* is a holopar. ther.

**holomycotroph** (often wrongly called saprophyte)
  e.g. *Thismia americana* is holomycotroph.

# Abbreviations

A ......................... alpine/arctic
Agg. ..................... aggregate
al. ........................ alii: others
Arch. .................. archipelago
app. .................... approaching, close to
auct. .................... of author
C. ........................ Central
cham. .................. chamaephyte
cit. ....................... citatus: cited
cl. ........................ climbing
Co. ...................... county
comb. .................. combinatio: combination
cons. .................... conservandus: to be conserved
cppo..................... centre page pull-out
cult. .................... cultus: cultivated
cv. ....................... cultivarietas: cultivar
descr. .................. description
Distr. .................. district
DT ....................... dry tropical (desert/steppe)
E. ........................ East(ern)
etc. ...................... et cetera: and the rest
e.g. ..................... exampli gratia: for example
G ........................ temperate
hel. ...................... helophyte
hort...................... hortorum: of gardens
I./Is ..................... island(s)
ICBN .................. International Code of Botanical Nomenclature
i.e. ...................... id est: that is
ign. ..................... gnotus: unknown
in litt. .................. in litteris: in correspondence
ined. ................... ineditus: unpublished, provisional name
inq. ..................... inquilinus: naturalised
i.q. ...................... idem quod: the same as
Kep. .................... kepulauan (islands)
Medit. ................. mediterranean
MT ...................... monsoon tropical (savanna)
Mt./Mts .............. mountain(s)
N. ....................... North(ern)
nanophan............. nanophanerophyte
No. ..................... numero: number
noh ..................... new orchid hybrids
nom. cons. .......... nomen conservandum: name conserved in ICBN
nom. illeg. .......... nomen illegitimum: illegitimate name

nom. inval. .......... invalid name
nom. nud. ............ nomen nudum: name without a description
nom. rejic. .......... nomen rejiciendum: name rejected in ICBN
nom. superfl. ....... nomen superfluum: name superfluous when published
nov. ..................... novus: new
orth. var. ............. orthographic variant
par. ..................... parasitic
Pen. .................... peninsula(r)
phan. .................. phanerophyte
p.p. ..................... pro parte: partly
Prov. ................... province
q.e. ..................... quod est: which is
q.v. ..................... quod vide: which see
Reg. .................... region
Rep. .................... republic
S .......................... Subtropical
S. ........................ South(ern)
seq. ..................... sequens: following
s.l. ...................... sensu lato: in the broad sense
sp. ....................... species
s.p. ..................... without page number
sphalm. ............... sphalmate: by mistake
s.s. ...................... sensu stricto: in the narrow sense
st. ....................... status
subtrop. .............. subtropical
syn. ..................... synonymon: synonym
T .......................... tropical
temp. .................. temperate
ther. .................... therophyte
trop. .................... tropical
vol. ..................... volume: volume
viz. ..................... videlicet: namely
W. ....................... West(ern)
WT ..................... wet tropical
? .......................... not known, doubtful
† .......................... extinct

# References

Bridson, G., comp. & ed. (2004). Botanico-Periodicum Huntianum/Second Edition. Pittsburgh: Hunt Institute for Botanical Documentation

Brummitt, R. K. (1992). Vascular Plant Families and Genera. 804 pp. Kew: Royal Botanic Gardens. [http://www.kew.org.uk/data/vascplnt.html]

Brummitt, R. K. (2001). World Geographical Scheme for Recording Plant Distributions, ed 2. xvi, 138 pp. Hunt Institute for Botanical Documentation, Carnegie-Mellon University, Pittsburgh, Penna. (for the International Working Group on Taxonomic Databases for Plant Sciences). (Plant Taxonomic Database Standards, 2: version 1.0.) [http://www.tdwg.org/geo2.htm]

Brummitt, R. K. & Powell, C. E., (1992). Authors of Plant Names. 732 pp. Kew: Royal Botanic Gardens. [http://www.ipni.org/ipni/query_author.html]

Burkill, I.H. (1951). Dioscoreaceae. In C. G. G. J van Steenis (ed.), *Flora Malesiana*, Ser. I, Vol. 4. pp. 293–335. Noordhoff-Kolff N.V., Djakarta.

Burkill, I.H. (1960). The organography and the evolution of the Dioscoreaceae, the family of the yams. *Journal of the Linnaean Society (Botany)* 56: 319–412.

Burkill, I.H. & Perrier de la Bâthie, H. (1950). Dioscoréacées, 44ᵉ famille. In H. Humbert (ed.), *Flore de Madagascar et des Comores*. Firmin-Didot et Cⁱᵉ, Paris.

Caddick, L.R, Rudall, P.J., Wilkin, P, Hedderson, T.A.J. & Chase, M.W. (2002a). Phylogenetics of Dioscoreales Based on Combined Analyses of Morphological and Molecular Data. *Botanical Journal of the Linnaean Society* 138: 123–144.

Caddick, L.R, Rudall, P.J., Wilkin, P, Hedderson, T.A.J. & Chase, M.W. (2002b). Yams reclassified: a recircumscription of Dioscoreaceae and Dioscoreales. *Taxon* 51: 103–114.

Coursey, D.G. (1967). *Yams: an account of the nature, origins, cultivation and utilisation of the useful members of the Dioscoreaceae.* Longmans, London.

De Langhe, J. E. et al. (1983). Flora van België, het Groothertogdom Luxemburg, Noord-Frankrijk en de aangrenzende gebieden. civ, 970 pp., illus., map. Patrimonium, Nationale Plantentuin van België, Meise.

Dounias, E. (2001). The management of wild yam tubers by the Baka pygmies in southern Cameroon. African Study Monographs, Suppl. 26: 135–156.

Dumont, R., Dansi, A., Vernier, P. & Zoundjikèkpon, J. (2006). Biodiversity and domestication of Yams in West Africa: traditional practices leading to *Dioscorea rotundata* Poir. CIRAD-IPGRI.

Farr, E. R., Leussink, J. A. & Stafleu, F. A. (eds). (1979). Index Nominum Genericorum (Plantarum). 3 vols. Bohn, Scheltema & Holkema, Utrecht. (Regnum Vegetabile, 100–102).

Greuter, W. et al. (1993). Names in Current Use for Extant Plant Genera (Names in current use, 3). xxvii, 1464 pp. Koeltz, Koenigstein. (Regnum Vegetabile.)

Hildebrand, E.A., Sebsebe D. & Wilkin, P. (2002). Local and regional landrace disappearance in species of *Dioscorea* (L.) (yams) in Southwest Ethiopia: causes of agrobiodiversity loss, and strategies for conservation. In Stepp, J.R.,Wyndham, F.S. & Zarger, R.K. (eds), *Ethnobiology and Biocultural Diversity*. pp. 678–695. University of Georgia Press, Athens, Georgia.

Hladik, A. & Dounias, E. (1996) Les ignames spontanées des forêts africaines, plantes à tubercules comestibles. In Hladik, C.M., Hladik, A., Pagezy, H., Linares, O., Koppert,

G.J.A. & Froment, A. (eds) *L'Alimentation en Forêt Tropicale : Interactions Bioculturelles et Perpesctives de Développement.* pp. 275–294. UNESCO, Paris.

IUCN. (2001). *IUCN Red List Categories.* Version 3.1. Prepared by the IUCN Species Survival Commission. IUCN, Gland, Switzerland & Cambridge, UK.

Jeannoda, Victor, Jeannoda, Vololoniaina, Hladik, A., & Hladik, C.M. (2003). Les ignames de Madagascar. Diversité, utilisations et perceptions. Hommes & Plantes 47: 10–23.

Jeannoda, V.H., Razanamparany, J.L., Rajanoah, M.T., Monneuse, M.O., Hladik, A. & Hladik, C.M. (2007). Les ignames (*Dioscorea* spp.) de Madagascar : espèces endémiques et formes introduites ; diversité, perception, valeur nutritionnelle et systèmes de gestion durable. Revue d'Ecologie (Terre Vie) 61 : in press.

Knuth, R. (1924). Dioscoreaceae. In Engler, H.G.A. (ed.) *Das Pflanzenreich*, 87 (IV. 43). pp. 1–387. Leipzig: H.R. Engelmann (J. Cramer).

Maas-van de Kamer, H. (1998). Burmanniaceae. In Kubitzki, K. (ed.), *The families and genera of vascular plants.* Volume III. Monocotyledons, Lilianae (except Orchidaceae). pp. 154–164. Springer-Verlag, Berlin.

Malapa, R., Arnau, G., Noyer, J.L. & Lebot, V. (2005). Genetic diversity of the greater yam (*Dioscorea alata* L.) and relatedness to *D. nummularia* Lam. and *D. transversa* R. Br. As revealed with AFLP markers. Genetic Resources & Crop Evolution 52: 919–929.

Marks, L. (2001). Sexual Chemistry: A History of the Contraceptive Pill. Yale University Press, New Haven, Connecticut.

Prain, D. & Burkill, I.H. (1936). An account of the genus *Dioscorea* in the East, Part 1: The species which twine to the left. *Annals of the Royal Botanic Gardens, Calcutta* 14: 1–210.

Prain, D. & Burkill, I.H. (1938). An account of the genus *Dioscorea* in the East, Part 2: The species which twine to the right. *Annals of the Royal Botanic Gardens, Calcutta* 14: 211–528.

Raunkiær, C. (1934). The Life Forms of Plants and Statistical Plant Geography. xvi, 632 pp., illus. Oxford University Press, London.

Raz, L. (2007). Systematics and biogeography of West Indian Dioscoreaceae. Unpubl. PhD Thesis, New York University, New York, USA.

Raz, L. (in prep.) New combinations plus two new species of West Indian Dioscoreaceae: Dioscorea sect. Rajania. To be submitted to Brittonia.

Sosa, V., Schubert, B.G & Gómez-Pompa, A. (1987). *Flora de Veracruz. Dioscoreaceae.* Instituto Nacional de Investigaciones Sobre Recursos Bioticos, Xalapa.

Sparg, S.G., Light, M.E. & van Staden, J. (2004). Biological activities and distribution of plant saponins. Journal of Ethnopharmacology 94: 219–243.

Stafleu, F. & Cowan, R. S. (1976–88). Taxonomic Literature: A Selective Guide to Botanical Publications and Collections with Dates, Commentaries and Types. 2nd edn. 7 vols. Utrecht: Bohn, Scheltema & Holkema. (Regnum Vegetabilie 94, 98, 105, 110, 112, 115, 116.) Continued as Stafleu, F. et al. (1992–2000). Taxonomic Literature, Supplement. Vols. 1–6. Koenigstein, Germany: Koeltz. (Regnum Vegetabile 125, passim. As of 2000 six volumes published.) [http://tl2.idcpublishers.info/]

Tamura, M. (1998). Nartheciaceae. In Kubitzki, K. (ed.), *The families and genera of vascular plants.* Volume III. Monocotyledons, Lilianae (except Orchidaceae).Pp. 381–392. Springer-Verlag, Berlin.

Tellez V., O & Schubert, V.G. (1994). Dioscoreaceae. In Davidse, G., Sousa S., M.S. & Chater, A.O. (eds), *Flora Mesoamericana* Vol. 6, Alismataceae a Cyperaceae. pp.54–65. Universidad Nacional Autonoma de Mexico, Mexico D.F.

Wilkin, P. (2001). Yams of South-Central Africa. Kew Bulletin 56: 361–404.

Wilkin, P. & Thapyai, C. (2007). *Flora of Thailand*: Dioscoreaceae. Submitted to Editorial Board.

Wilkin, P., Hladik, A., Labat, J.-N. & Barthelat, F. (2007a). A new edible yam (*Dioscorea* L.) species endemic to Mayotte, new data on *D. comorensis* R. Knuth and a key to the yams of the Comoro Archipelago. Adansonia XX: submitted.

Wilkin, P., Rajaonah, M.T., Jeannoda, V.H., Hladik, A. Jeannoda V. & Hladik, C.M. (2007b). An endangered new species of edible yam (*Dioscorea, Dioscoreaceae*) from Western Madagascar. Kew Bulletin XX: submitted.

Wilkin, P., Weber, O. & Moat, J. Yams (Dioscorea L.) of Madagascar: systematics and conservation status. MS in prep.

# Acknowledgements

In particular we would like to thank Nick Black for writing and maintaining the computer programmes and reports used to create the checklist. Thanks are also due to those who have participated in or facilitated taxonomic study of Dioscoreales: Liz Caddick, Chirdsak Thapyai, Anna Haigh, Odile Weber, Kate Davis, Annette Hladik, Vololoniaiana Jeannoda, Jean-Noel Labat, Fabien Barthelat, Thierry Deroin, Mark Newman, Kongkanda Chayamarit, Franck Rakotonasolo, COLPARSYST, SYNTHESYS, the QBG-DANCED programme, the John Spedan Lewis Trust, and the herbarium curators cited in the original publications.

# Burmanniaceae

## Afrothismia

*Afrothismia* Schltr., Bot. Jahrb. Syst. 38: 138 (1906).
WC. & E. Trop. Africa. 22 NGA 23 CMN 25 KEN
TAN UGA.
10 Species

*Afrothismia baerae* Cheek, Kew Bull. 58: 951 (2003
publ. 2004).
Kenya. 25 KEN. Holomycotrophic rhizome geophyte.

*Afrothismia foertheriana* T.Franke, Sainge & Agerer,
Blumea 49: 452 (2004).
Cameroon. 23 CMN. Holomycotrophic rhizome
geophyte.

*Afrothismia gesnerioides* H.Maas, Blumea 48: 477
(2003).
Cameroon. 23 CMN. Holomycotrophic rhizome
geophyte.

*Afrothismia hydra* Sainge & T.Franke, Nordic J. Bot. 23:
299 (2005).
Cameroon. 23 CMN. Holomycotrophic rhizome
geophyte.

*Afrothismia insignis* Cowley, in Fl. Trop. E. Afr.,
Burmann.: 7 (1988).
SW. Tanzania. 25 TAN. Holomycotrophic rhizome
geophyte.

*Afrothismia korupensis* Sainge & T.Franke, Willdenowia
35: 289 (2005).
SW. Cameroon. 23 CMN. Holomycotrophic rhizome
geophyte.

*Afrothismia mhoroana* Cheek, Kew Bull. 60: 593 (2005
publ. 2006).
Tanzania (Morogoro Distr.). 25 TAN.
Holomycotrophic rhizome geophyte.

*Afrothismia pachyantha* Schltr., Bot. Jahrb. Syst. 38:
139 (1906). *Thismia pachyantha* (Schltr.) Engl. in
H.G.A.Engler & C.G.O.Drude, Veg. Erde 9(2): 401
(1908).
Cameroon. 23 CMN. Holomycotrophic rhizome
geophyte.

*Afrothismia saingei* T.Franke, Syst. Geogr. Pl. 74: 28
(2004).
Cameroon (Mt. Kupe). 23 CMN. Holomycotrophic
rhizome geophyte.

*Afrothismia winkleri* (Engl.) Schltr., Bot. Jahrb. Syst.
38: 139 (1906).
S. Nigeria to SW. Uganda. 22 NGA 23 CMN 25 UGA.
Holomycotrophic rhizome geophyte.
*Thismia winkleri* Engl., Bot. Jahrb. Syst. 38: 89 (1905).

var. *budongensis* Cowley, in Fl. Trop. E. Afr.,
Burmann.: 7 (1988).
SW. Uganda. 25 UGA. Holomycotrophic rhizome
geophyte.

var. *winkleri*
S. Nigeria to Cameroon. 22 NGA 23 CMN.
Holomycotrophic rhizome geophyte.

## Apteria

*Apteria* Nutt., J. Acad. Nat. Sci. Philadelphia 7: 64 (1834).
Trop. & Subtrop. America. 77 TEX 78 ALA FLA
GEO LOU MSI 79 MXC MXG MXS MXT 80
BLZ COS GUA HON NIC PAN 81 CUB DOM
HAI JAM LEE PUE TRT WIN 82 FRG GUY SUR
VEN 83 BOL CLM ECU PER 84 BZC BZE BZL
BZN BZS 85 PAR.
1 Species
*Nemitis* Raf., Fl. Tellur. 4: 33 (1838).
*Stemoptera* Miers, Proc. Linn. Soc. London 1: 62
(1840).

*Apteria aphylla* (Nutt.) Barnhart ex Small, Fl. S.E. U.S.:
309 (1903).
Trop. & Subtrop. America. 77 TEX 78 ALA FLA GEO
LOU MSI 79 MXC MXG MXS MXT 80 BLZ COS
GUA HON NIC PAN 81 CUB DOM HAI JAM
LEE PUE TRT WIN 82 FRG GUY SUR VEN 83
BOL CLM ECU PER 84 BZC BZE BZL BZN BZS
85 PAR. Holomycotrophic rhizome geophyte.
*Apteria setacea* Nutt., J. Acad. Nat. Sci. Philadelphia 7:
64 (1834). *Nemitis setacea* (Nutt.) Raf., Fl. Tellur.
4: 33 (1838).
*Stemoptera lilacina* Miers, Proc. Linn. Soc. London 1:
62 (1840). *Apteria lilacina* (Miers) Miers, Trans.
Linn. Soc. London 18: 546 (1841).
*Apteria setacea* var. *major* Hook., Hooker's Icon. Pl. 7:
t. 660 (1844).
*Apteria hymenanthera* Miq., Stirp. Surinam. Select.: 216
(1851). *Apteria aphylla* var. *hymenanthera* (Miq.)
Jonker in A.A.Pulle, Fl. Suriname 1(1): 186 (1938).
*Apteria ulei* Schltr., Verh. Bot. Vereins Prov.
Brandenburg 47: 102 (1905).
*Apteria boliviana* Rusby, Bull. New York Bot. Gard. 4:
447 (1907).
*Apteria gentianoides* Jonker, Monogr. Burmann.: 211
(1938).
*Apteria hymenanthera* f. *decolorata* Cif., Atti Ist. Bot.
Lab. Crittog. Univ. Pavia, V, 7: 27 (1946).

**Synonyms:**
*Apteria aphylla* var. *hymenanthera* (Miq.) Jonker =
**Apteria aphylla** (Nutt.) Barnhart ex Small
*Apteria boliviana* Rusby = **Apteria aphylla** (Nutt.)
Barnhart ex Small
*Apteria gentianoides* Jonker = **Apteria aphylla** (Nutt.)
Barnhart ex Small
*Apteria hymenanthera* Miq. = **Apteria aphylla** (Nutt.)
Barnhart ex Small
*Apteria hymenanthera* f. *decolorata* Cif. = **Apteria
aphylla** (Nutt.) Barnhart ex Small
*Apteria lilacina* (Miers) Miers = **Apteria aphylla** (Nutt.)
Barnhart ex Small
*Apteria orobanchoides* Hook. = **Dictyostega
orobanchoides** (Hook.) Miers
*Apteria setacea* Nutt. = **Apteria aphylla** (Nutt.) Barnhart
ex Small
*Apteria setacea* var. *major* Hook. = **Apteria aphylla**
(Nutt.) Barnhart ex Small
*Apteria ulei* Schltr. = **Apteria aphylla** (Nutt.) Barnhart ex
Small

## Bagnisia

*Bagnisia* Becc. = *Thismia* Griff.
*Bagnisia crocea* Becc. = *Thismia crocea* (Becc.) J.J.Sm.
*Bagnisia episcopalis* (Becc.) Engl. = *Thismia episcopalis* (Becc.) F.Muell.
*Bagnisia hillii* Cheeseman = *Thismia rodwayi* F.Muell.
*Bagnisia rodwayi* (F.Muell.) F.Muell. = *Thismia rodwayi* F.Muell.

## Benitzia

*Benitzia* H.Karst. = *Gymnosiphon* Blume
*Benitzia poeppigiana* H.Karst. = *Gymnosiphon divaricatus* (Benth.) Benth. & Hook.f.
*Benitzia suaveolens* H.Karst. = *Gymnosiphon suaveolens* (H.Karst.) Urb.

## Burmannia

*Burmannia* L., Sp. Pl.: 287 (1753).
Trop. & Subtrop. 22 GHA GUI IVO LBR NGA SEN SIE 23 BUR CAF CMN CON GAB RWA ZAI 24 CHA SUD 25 TAN UGA 26 ANG MOZ ZAM ZIM 27 BOT NAT TVL 29 MAU MDG 36 CHC CHH CHS 38 JAP NNS TAI 40 ASS BAN EHM IND NEP SRL 41 CBD LAO MYA THA VIE 42 BOR JAW LSI MLY MOL PHI SUL SUM 43 BIS NWG 50 NSW NTA QLD WAU 62 CRL 74 OKL 77 TEX 78 ALA FLA GEO LOU MSI NCA SCA VRG 79 MXG MXT 80 BLZ COS HON NIC PAN 81 CUB DOM JAM PUE TRT 82 FRG GUY SUR VEN 83 BOL CLM ECU PER 84 BZC BZE BZL BZN BZS 85 AGE PAR.
57 Species
*Vogelia* J.F.Gmel., Syst. Nat. 2: 107 (1791).
*Tripterella* Michx., Fl. Bor.-Amer. 1: 19 (1803).
*Maburnia* Thouars, Gen. Nov. Madagasc.: 4 (1806).
*Gonianthes* Blume, Catalogus: 19 (1823).
*Gonyanthes* Nees, Ann. Sci. Nat. (Paris) 3: 369 (1824), orth. var.
*Tetraptera* Miers in J.Lindley, Veg. Kingd., ed. 2.: 172 (1847).
*Tripteranthus* Wall. ex Miers in J.Lindley, Veg. Kingd., ed. 2.: 172 (1847).
*Cryptonema* Turcz., Bull. Soc. Imp. Naturalistes Moscou 21(1): 590 (1848).
*Nephrocoelium* Turcz., Bull. Soc. Imp. Naturalistes Moscou 1853(1): 287 (1853).

*Burmannia alba* Mart., Nov. Gen. Sp. Pl. 1: 12 (1824). *Tripterella alba* (Mart.) Schult. in J.J.Roemer & J.A.Schultes, Mant. 1: 358 (1822).
Brazil to Paraguay. 84 BZC BZE BZL BZS 85 PAR. Holopar. tuber geophyte.
*Burmannia sellowiana* Seub. in C.F.P.von Martius & auct. suc. (eds.), Fl. Bras. 3(1): 57 (1847).
*Burmannia sellowiana* var. *albiflora* Seub. in C.F.P.von Martius & auct. suc. (eds.), Fl. Bras. 3(1): 57 (1847).
*Burmannia sellowiana* var. *violacea* Seub. in C.F.P.von Martius & auct. suc. (eds.), Fl. Bras. 3(1): 57 (1847).

*Burmannia aprica* (Malme) Jonker, Monogr. Burmann.: 87 (1938).
SE. Brazil. 84 BZL.
*\*Burmannia bicolor* var. *aprica* Malme, Bih. Kongl. Svenska Vetensk.-Akad. Handl. 22(3, 8): 22 (1896).
*Burmannia aprica* var. *pusilla* Jonker, Monogr. Burmann.: 88 (1938).

*Burmannia australis* Malme, Bih. Kongl. Svenska Vetensk.-Akad. Handl. 22(3, 8): 25 (1896).
Brazil, Bolivia, Paraguay. 83 BOL 84 BZE BZL BZS 85 PAR. Ther.

*Burmannia bicolor* Mart., Nov. Gen. Sp. Pl. 1: 10 (1824). *Tripterella bicolor* (Mart.) Schult. in J.J.Roemer & J.A.Schultes, Mant. 1: 357 (1822).
Cuba, S. Trop. America. 81 CUB 82 FRG GUY SUR VEN 83 CLM 84 BZC BZE BZL BZN.
*Burmannia brachyphylla* Willd. ex Schult. & Schult.f. in J.J.Roemer & J.A.Schultes, Syst. Veg. 7(2): lxxv (1830).
*Burmannia quadriflora* Willd. ex Schult. & Schult.f. in J.J.Roemer & J.A.Schultes, Syst. Veg. 7(2): lxxiv, 741 (1830). *Burmannia bicolor* var. *quadriflora* (Willd. ex Schult. & Schult.f.) Malme, Bot. Not. 1898: 186 (1898).
*Burmannia brachystachya* Miq., Linnaea 19: 141 (1846).
*Burmannia bicolor* var. *subcoelestis* Malme, Bih. Kongl. Svenska Vetensk.-Akad. Handl. 22(3, 8): 21 (1896).

*Burmannia bifaria* J.J.Sm., Icon. Bogor. 4: t. 379 (1914).
W. Jawa. 42 JAW. Holopar. geophyte.

*Burmannia biflora* L., Sp. Pl.: 287 (1753). *Tripterella caerulea* Muhl. ex Nutt., Gen. N. Amer. Pl. 1: 22 (1818), nom. illeg. *Tripterella biflora* (L.) Schult. in J.J.Roemer & J.A.Schultes, Mant. 1: 356 (1822).
SE. Europe to Texas, W. Cuba. 77 TEX 78 ALA FLA GEO LOU MSI NCA SCA VRG 81 CUB. Ther.

*Burmannia candelabrum* Gagnep., Bull. Soc. Bot. France 54: 462 (1907).
India, Assam, Bangladesh. 40 ASS BAN IND. Ther.

*Burmannia candida* Griff. ex Hook.f., Fl. Brit. India 5: 665 (1888).
Indo-China, Sumatera. 41 MYA THA 42 SUM.

*Burmannia capitata* (Walter ex J.F.Gmel.) Mart., Nov. Gen. Sp. Pl. 1: 12 (1824).
Trop. & Subtrop. America. 74 OKL 77 TEX 78 ALA FLA GEO LOU MSI NCA SCA 79 MXG MXT 80 BLZ COS HON NIC PAN 81 CUB DOM JAM PUE TRT 82 FRG GUY SUR VEN 83 BOL CLM 84 BZC BZE BZL BZN BZS 85 AGE PAR. Ther.
*\*Vogelia capitata* Walter ex J.F.Gmel., Syst. Nat. 2: 107 (1791) (1791). *Tripterella capitata* (Walter ex J.F.Gmel.) Michx., Fl. Bor.-Amer. 1: 19 (1803). *Gyrotheca capitata* (Walter ex J.F.Gmel.) Morong, Bull. Torrey Bot. Club 20: 472 (1893).
*Burmannia bracteosa* Gleason, Bull. Torrey Bot. Club 58: 343 (1931). *Burmannia capitata* f. *bracteosa* (Gleason) Jonker, Monogr. Burmann.: 74 (1938).

*Burmannia championii* Thwaites, Enum. Pl. Zeyl.: 325 (1864).
Trop. Asia to C. Japan. 36 CHS 38 JAP TAI 40 SRL 42 BOR JAW MOL SUM 43 NWG. Holomycotrophic ther.
*Burmannia tuberosa* Becc., Malesia 1: 245 (1878).
*Burmannia dalzielii* Rendle, J. Bot. 40: 311 (1902).
*Burmannia chionantha* Schltr., Bot. Jahrb. Syst. 49: 107 (1912).
*Burmannia japonica* Maxim. ex Matsum., Index Pl. Jap. 2: 234 (1912).
*Burmannia hunanensis* K.M.Liu & C.L.Long, Ann. Bot. Fenn. 38: 211 (2001).

*Burmannia chinensis* Gand., Bull. Soc. Bot. France 66: 290 (1919 publ. 1920).

E. India, S. China to Indo-China, Nansei-shoto (Iriomate). 36 CHC CHH CHS 38 NNS 40 IND 41 LAO THA VIE. Holomycotrophic ther.
*Burmannia rigida* Gand., Bull. Soc. Bot. France 66: 290 (1919 publ. 1920).
*Burmannia urazii* Masam., Trans. Nat. Hist. Soc. Taiwan 24: 207 (1934).
*Burmannia pusilla* var. *hongkongensis* Jonker, Monogr. Burmann.: 131 (1938).

**Burmannia cochinchinensis** Gagnep., Bull. Soc. Bot. France 54: 463 (1907).
N. Vietnam. 41 VIE. Ther.

**Burmannia coelestis** D.Don, Prodr. Fl. Nepal.: 44 (1825).
Nepal to Caroline Is. 36 CHH CHS 40 ASS BAN EHM IND NEP 41 CBD LAO MYA THA VIE 42 BOR JAW LSI MLY MOL PHI SUL SUM 43 NWG 50 NTA 62 CRL. Holomycotrophic ther.
*Burmannia javanica* Blume, Enum. Pl. Javae 1: 28 (1827).
*Burmannia uniflora* Rottler ex Spreng., Syst. Veg. 4(2): 23 (1827).
*Burmannia triflora* Roxb., Fl. Ind. ed. 1832, 2: 117 (1832).
*Cryptonema malaccensis* Turcz., Bull. Soc. Imp. Naturalistes Moscou 21(1): 591 (1848). *Nephrocoelium malaccensis* (Turcz.) Turcz., Bull. Soc. Imp. Naturalistcs Moscou 1853(1): 287 (1853).
*Burmannia azurea* Griff., Not. Pl. Asiat. 3: 236 (1851).
*Burmannia selebica* Becc., Malesia 1: 243 (1878).
*Burmannia bifurca* Ham. ex Hook.f., Fl. Brit. India 5: 665 (1888).
*Burmannia candida* var. *caerulea* Hook.f. ex F.N.Williams, Bull. Herb. Boissier, II, 4: 362 (1904).
*Burmannia borneensis* Gand., Bull. Soc. Bot. France 66: 290 (1919 publ. 1920).
*Burmannia malaccensis* Gand., Bull. Soc. Bot. France 66: 290 (1919 publ. 1920).

**Burmannia compacta** Maas & H.Maas, Fl. Neotrop. Monogr. 42: 53 (1986).
Venezuela (Amazonas). 82 VEN. Hemicr. or cham.

**Burmannia congesta** (C.H.Wright) Jonker, Monogr. Burmann.: 94 (1938).
W. Trop. Africa to Angola. 22 GHA LBR NGA 23 CMN GAB ZAI 26 ANG. Holomycotrophic ther.
*Gymnosiphon congestus* C.H.Wright in D.Oliver & auct. suc. (eds.), Fl. Trop. Afr. 7: 12 (1897).
*Burmannia aptera* Schltr., Bot. Jahrb. Syst. 38: 141 (1906).
*Burmannia densiflora* Schltr., Bot. Jahrb. Syst. 38: 141 (1906).

**Burmannia connata** Jonker, Monogr. Burmann.: 128 (1938).
E. Sumatera. 42 SUM. Ther.

**Burmannia cryptopetala** Makino, Bot. Mag. (Tokyo) 27: 3 (1913).
China, Japan (S. Honshu, Kyushu). 36 CHH CHS 38 JAP.
*Burmannia cryptopetala* var *daxikangensis* Y.B.Chang & Z.Wei, Bull. Bot. Res., Harbin 9(2): 37 (1989).
*Burmannia daxikangensis* (Y.B.Chang & Z.Wei) S.C.Chen & H.Li, Acta Bot. Yunnan. 27: 248 (2005).

**Burmannia damazii** Beauverd, Bull. Herb. Boissier, II, 5: 948, 1081 (1905).
C. & SE. Brazil. 84 BZC BZL.

*Burmannia dasyantha* Mart., Nov. Gen. Sp. Pl. 1: 11 (1824). *Tripterella dasyantha* (Mart.) Schult. in J.J.Roemer & J.A.Schultes, Mant. 1: 357 (1822).
SE. Colombia to SW. Venezuela. 82 VEN 83 CLM.

**Burmannia disticha** L., Sp. Pl.: 287 (1753).
S. China to Trop. Asia, E. Australia. 36 CHC CHS 40 EHM IND NEP SRL 41 VIE 42 BOR SUL SUM 43 NWG 50 NSW QLD. Ther.
*Burmannia distachya* R.Br., Prodr.: 265 (1810), nom. illeg.
*Burmannia bancana* Miq., Fl. Ned. Ind., Eerste Bijv. 1: 617 (1860).
*Burmannia sumatrana* Miq., Fl. Ned. Ind., Eerste Bijv. 1: 616 (1860). *Burmannia disticha* var. *sumatrana* (Miq.) Hook.f., Fl. Brit. India 5: 664 (1888).
*Burmannia graminifolia* Warb., Repert. Spec. Nov. Regni Veg. 18: 330 (1922).

**Burmannia engganensis** Jonker, Blumea 3: 108 (1938).
Sumatera (Enggano I.). 42 SUM. Holopar. geophyte.

**Burmannia filamentosa** D.X.Zhang & R.M.K.Saunders, Nordic J. Bot. 20: 392 (2000 publ. 2001).
China (Guangdong). 36 CHS.

**Burmannia flava** Mart., Nov. Gen. Sp. Pl. 1: 11 (1824). *Tripterella flava* (Mart.) Schult. in J.J.Roemer & J.A. Schultes, Mant. 1: 257 (1822).
Trop. & Subtrop. America. 78 FLA 79 MXT 80 BLZ COS PAN 81 CUB 82 FRG GUY SUR VEN 83 BOL CLM 84 BZC BZL BZN BZS 85 AGE PAR. Ther.
*Burmannia flavula* C.Wright in F.A.Sauvalle, Fl. Cub.: 165 (1871).
*Burmannia flava* var. *macroptera* Jonker, Monogr. Burmann.: 85 (1938).

**Burmannia foliosa** Gleason, Bull. Torrey Bot. Club 58: 343 (1931).
S. Venezuela. 82 VEN. Hemicr.
*Burmannia foliosa* subsp. *marahuacensis* Maguire & Steyerm., Acta Bot. Venez. 14: 19 (1984).

**Burmannia geelvinkiana** Becc., Malesia 1: 244 (1878).
W. New Guinea. 43 NWG. Ther.

**Burmannia gracilis** Ridl., J. Straits Branch Roy. Asiat. Soc. 22: 335 (1890).
Pen. Thailand to Pen. Malaysia. 41 THA 42 MLY. Holomycotroph.

**Burmannia grandiflora** Malme, Bih. Kongl. Svenska Vetensk.-Akad. Handl. 22(3, 8): 27 (1896). *Burmannia alba* var. *grandiflora* (Malme) Jonker, Monogr. Burmann.: 76 (1938).
Colombia, C. Brazil. 83 CLM 84 BZC. Holomycotrophic tuber geophyte.

**Burmannia hexaptera** Schltr., Bot. Jahrb. Syst. 38: 143 (1906).
Cameroon to Gabon. 23 CMN GAB. Ther.

**Burmannia indica** Jonker, Monogr. Burmann.: 161 (1938).
S. India. 40 IND.

**Burmannia itoana** Makino, Bot. Mag. (Tokyo) 27: 1 (1913).
S. China to Japan (Yakushima). 36 CHC CHH CHS 38 JAP NNS TAI. Holomycotrophic rhizome geophyte.
*Burmannia takeoi* Hayata, Icon. Pl. Formosan. 5: 212 (1915).

*Burmannia pingbienensis* H.Li, Acta Phytotax. Sin. 21: 127 (1983).

**Burmannia jonkeri** Benthem & Maas, Acta Bot. Neerl. 30: 140 (1981).
Brazil (Mato Grosso, Goiás). 84 BZC. Holomycotrophic tuber geophyte.

**Burmannia juncea** Sol. ex R.Br., Prodr.: 265 (1810).
N. Australia. 50 NTA QLD WAU.

**Burmannia kalbreyeri** Oliv., Hooker's Icon. Pl. 14: t. 1357 (1881).
Costa Rica to Venezuela and Peru. 80 COS PAN 82 VEN 83 CLM ECU PER. Hemicr. or epiphyte.
*Burmannia wercklei* Schltr., Repert. Spec. Nov. Regni Veg. 22: 35 (1913).
*Burmannia herthae* G.M.Schulze, Notizbl. Bot. Gart. Berlin-Dahlem 15: 43 (1940).
*Burmannia carrenoi* Steyerm., Acta Bot. Venez. 6: 89 (1971 publ. 1972).

**Burmannia larseniana** D.X.Zhang & R.M.K.Saunders, Nordic J. Bot. 19: 243 (1999).
Thailand. 41 THA.

**Burmannia latialata** Pobég., Essai Fl. Guinée Franç.: 166 (1906).
W. Trop. Africa to Uganda. 22 GHA GUI IVO LBR NGA SEN SIE 23 CAF CMN CON GAB ZAI 24 CHA SUD 25 UGA. Ther.
*Burmannia bicolor* var. *africana* Ridl., J. Bot. 25: 85 (1887).
*Burmannia bicolor* var. *micrantha* Engl. & Gilg in O.Warburg (ed.), Kunene-Sambesi Exped.: 202 (1903).
*Burmannia liberica* Engl., Bot. Jahrb. Syst. 48: 505 (1912).
*Burmannia chariensis* Schltr., Repert. Spec. Nov. Regni Veg. 21: 82 (1925).
*Burmannia le-testui* Schltr., Repert. Spec. Nov. Regni Veg. 21: 82 (1925).
*Burmannia obscurata* Schltr., Repert. Spec. Nov. Regni Veg. 21: 83 (1925).
*Burmannia welwitschii* Schltr., Repert. Spec. Nov. Regni Veg. 21: 84 (1925).

**Burmannia ledermannii** Jonker, Monogr. Burmann.: 126 (1938).
New Guinea, Caroline Is. (Palau). 43 NWG 62 CRL.

**Burmannia longifolia** Becc., Malesia 1: 244 (1878).
Malesia to Papuasia. 42 BOR MLY MOL PHI SUL SUM 43 BIS NWG.
*Burmannia leucantha* Schltr., Bot. Jahrb. Syst. 49: 107 (1912).

**Burmannia luteoalba** Gagnep., Bull. Soc. Bot. France 54: 463 (1907).
Cambodia (Phu-quoc I.), Vietnam. 41 CBD VIE. Ther.

**Burmannia lutescens** Becc., Malesia 1: 246 (1878).
Malesia to Papuasia. 42 BOR JAW MLY SUL SUM 43 BIS NWG. Holomycotroph.
*Gonianthes candida* Blume, Catalogus: 20 (1823).
*Burmannia candida* (Blume) Engl. in H.G.A.Engler & K.A.E.Prantl, Nat. Pflanzenfam. 2(6): 50 (1888), nom. illeg.
*Burmannia tridentata* Becc., Malesia 1: 246 (1878).
*Burmannia papillosa* Stapf, Trans. Linn. Soc. London, Bot. 4: 232 (1894).
*Burmannia novae-hiberniae* Schltr. in K.M.Schumann & C.A.G.Lauterbach, Fl. Schutzgeb. Südsee, Nachtr.: 73 (1905).

*Burmannia gjellerupii* J.J.Sm., Repert. Spec. Nov. Regni Veg. 10: 487 (1912).
*Burmannia gonyantha* Hochr., Candollea 2: 325 (1925).

**Burmannia madagascariensis** Mart., Nov. Gen. Sp. Pl. 1: 12 (1824).
Rwanda to S. Africa, Mauritius, Madagascar. 23 BUR RWA ZAI 25 TAN 26 ANG MOZ ZAM ZIM 27 BOT NAT TVL 29 MAU MDG. Ther.
*Burmannia capensis* Mart., Nov. Gen. Sp. Pl. 1: 12 (1824).
*Burmannia paniculata* Willd. ex Schult. & Schult.f. in J.J.Roemer & J.A.Schultes, Syst. Veg. 7(2): lxxiv (1830).
*Burmannia madagascariensis* Baker, J. Linn. Soc., Bot. 20: 268 (1883), nom. illeg.
*Burmannia blanda* Gilg in O.Warburg (ed.), Kunene-Sambesi Exped.: 202 (1903).
*Burmannia bakeri* Hochr., Annuaire Conserv. Jard. Bot. Genève 11-12: 54 (1908).
*Burmannia inhambanensis* Schltr., Repert. Spec. Nov. Regni Veg. 11: 82 (1912).

**Burmannia malasica** Jonker, Monogr. Burmann.: 152 (1938).
Pen. Thailand, SE. Borneo. 41 THA 42 BOR.

**Burmannia micropetala** Ridl., Trans. Linn. Soc. London, Bot. 9: 228 (1916).
New Guinea. 43 NWG. Holomycotroph.

**Burmannia nepalensis** (Miers) Hook.f., Fl. Brit. India 5: 666 (1888).
Nepal to Philippines and Japan (Tanegashima). 36 CHC CHS 38 JAP NNS TAI 40 ASS NEP 41 VIE 42 PHI.
*Gonianthes nepalensis* Miers, Trans. Linn. Soc. London 18: 537 (1841). *Cyanotis nepalensis* (Miers) Miers in N.Wallich, Numer. List: 9006 (1849).
*Burmannia clementis* Schltr., Philipp. J. Sci. 1(Suppl.): 305 (1906).
*Burmannia liukiuensis* Hayata, Icon. Pl. Formosan. 5: 211 (1915).
*Burmannia fadouensis* H.Li, Acta Phytotax. Sin. 21: 125 (1983).

**Burmannia oblonga** Ridl., J. Straits Branch Roy. Asiat. Soc. 41: 33 (1904).
Hainan, Vietnam, Pen. Malaysia, N. Sumatera. 36 CHH 41 VIE 42 MLY SUM. Holomycotroph.
*Burmannia bifida* Gagnep., Bull. Soc. Bot. France 54: 462 (1907).

**Burmannia polygaloides** Schltr., Verh. Bot. Vereins Prov. Brandenburg 47: 103 (1905).
S. Venezuela to W. & C. Brazil. 82 VEN 84 BZC BZN. Hemicr. or cham.

**Burmannia pusilla** (Miers) Thwaites, Enum. Pl. Zeyl.: 325 (1864).
India, Assam, Sri Lanka, Indo-China. 40 ASS IND SRL 41 CBD VIE.
*Gonianthes pusilla* Miers, Trans. Linn. Soc. London 18: 537 (1841). *Cyananthus pusillus* (Miers) Miers in N.Wallich, Numer. List: 9008 (1849). *Burmannia coelestis* var. *pusilla* (Miers) Trimen, Handb. Fl. Ceylon 4: 131 (1898).

**Burmannia sanariapoana** Steyerm., Fieldiana, Bot. 28: 165 (1951).
S. Venezuela. 82 VEN.

**Burmannia sphagnoides** Becc., Malesia 1: 246 (1878).
Pen. Malaysia, Sumatera, W. Borneo. 42 BOR MLY SUM. Holomycotroph.

*Burmannia steenisii* Jonker, Monogr. Burmann.: 158 (1938).
E. Jawa (Mt. Lamongan). 42 JAW. Holomycotroph.

*Burmannia stricta* Jonker, Monogr. Burmann.: 156 (1938).
S. India. 40 IND.

*Burmannia stuebelii* Hieron. & Schltr., Bot. Jahrb. Syst. 54(117): 15 (1916).
N. Peru. 83 PER.

*Burmannia subcoelestis* Gagnep., Bull. Soc. Bot. France 54: 464 (1907).
Indo-China. 41 CBD LAO VIE. Ther.

*Burmannia tenella* Benth., Hooker's J. Bot. Kew Gard. Misc. 7: 12 (1855).
S. Trop. America. 82 GUY VEN 83 BOL CLM ECU PER 84 BZC BZN. Holomycotrophic hemicr.
*Burmannia amazonica* Schltr., Verh. Bot. Vereins Prov. Brandenburg 47: 102 (1905).

*Burmannia tenera* (Malme) Jonker, Monogr. Burmann.: 86 (1938).
Brazil (Goiás, São Paulo). 84 BZC BZL.
*\*Burmannia bicolor* var. *tenera* Malme, Bih. Kongl. Svenska Vetensk.-Akad. Handl. 22(3, 8): 23 (1896).

*Burmannia tisserantii* Schltr., Repert. Spec. Nov. Regni Veg. 21: 84 (1925).
Central African Rep. 23 CAF. Ther.

*Burmannia vaupesiana* Benthem & Maas, Acta Bot. Neerl. 30: 141 (1981).
Colombia. 83 CLM.

*Burmannia wallichii* (Miers) Hook.f., Fl. Brit. India 5: 666 (1888).
India to SE. China and Pen. Malaysia. 36 CHH CHS 40 IND 41 CBD MYA THA VIE 42 MLY. Holopar. ther.
*\*Gonianthes wallichii* Miers, Trans. Linn. Soc. London 18: 537 (1841).
*Burmannia griffithii* Becc., Malesia 1: 254 (1878).

*Synonyms*:
*Burmannia alba* var. *grandiflora* (Malme) Jonker = *Burmannia grandiflora* Malme
*Burmannia amazonica* Schltr. = *Burmannia tenella* Benth.
*Burmannia aprica* var. *pusilla* Jonker = *Burmannia aprica* (Malme) Jonker
*Burmannia aptera* Schltr. = *Burmannia congesta* (C.H.Wright) Jonker
*Burmannia azurea* Griff. = *Burmannia coelestis* D.Don
*Burmannia bakeri* Hochr. = *Burmannia madagascariensis* Mart.
*Burmannia bancana* Miq. = *Burmannia disticha* L.
*Burmannia bicolor* var. *africana* Ridl. = *Burmannia latialata* Pobég.
*Burmannia bicolor* var. *aprica* Malme = *Burmannia aprica* (Malme) Jonker
*Burmannia bicolor* var. *micrantha* Engl. & Gilg = *Burmannia latialata* Pobég.
*Burmannia bicolor* var. *quadriflora* (Willd. ex Schult. & Schult.f.) Malme = *Burmannia bicolor* Mart.
*Burmannia bicolor* var. *subcoelestis* Malme = *Burmannia bicolor* Mart.
*Burmannia bicolor* var. *tenera* Malme = *Burmannia tenera* (Malme) Jonker
*Burmannia bifida* Gagnep. = *Burmannia oblonga* Ridl.
*Burmannia bifurca* Ham. ex Hook.f. = *Burmannia coelestis* D.Don

*Burmannia blanda* Gilg = *Burmannia madagascariensis* Mart.
*Burmannia borneensis* Gand. = *Burmannia coelestis* D.Don
*Burmannia brachyphylla* Willd. ex Schult. & Schult.f. = *Burmannia bicolor* Mart.
*Burmannia brachystachya* Miq. = *Burmannia bicolor* Mart.
*Burmannia bracteosa* Gleason = *Burmannia capitata* (Walter ex J.F.Gmel.) Mart.
*Burmannia candida* (Blume) Engl. = *Burmannia lutescens* Becc.
*Burmannia candida* var. *caerulea* Hook.f. ex F.N.Williams = *Burmannia coelestis* D.Don
*Burmannia capensis* Mart. = *Burmannia madagascariensis* Mart.
*Burmannia capitata* f. *bracteosa* (Gleason) Jonker = *Burmannia capitata* (Walter ex J.F.Gmel.) Mart.
*Burmannia carrenoi* Steyerm. = *Burmannia kalbreyeri* Oliv.
*Burmannia chariensis* Schltr. = *Burmannia latialata* Pobég.
*Burmannia chionantha* Schltr. = *Burmannia championii* Thwaites
*Burmannia clementis* Schltr. = *Burmannia nepalensis* (Miers) Hook.f.
*Burmannia coelestis* var. *pusilla* (Miers) Trimen = *Burmannia pusilla* (Miers) Thwaites
*Burmannia cryptopetala* var. *daxikangensis* Y.B.Chang & Z.Wei = *Burmannia cryptopetala* Makino
*Burmannia dalzielii* Rendle = *Burmannia championii* Thwaites
*Burmannia densiflora* Schltr. = *Burmannia congesta* (C.H.Wright) Jonker
*Burmannia distachya* R.Br. = *Burmannia disticha* L.
*Burmannia disticha* var. *sumatrana* (Miq.) Hook.f. = *Burmannia disticha* L.
*Burmannia fadouensis* H.Li = *Burmannia nepalensis* (Miers) Hook.f.
*Burmannia flava* var. *macroptera* Jonker = *Burmannia flava* Mart.
*Burmannia flavula* C.Wright = *Burmannia flava* Mart.
*Burmannia foliosa* subsp. *marahuacensis* Maguire & Steyerm. = *Burmannia foliosa* Gleason
*Burmannia gjellerupii* J.J.Sm. = *Burmannia lutescens* Becc.
*Burmannia gonyantha* Hochr. = *Burmannia lutescens* Becc.
*Burmannia graminifolia* Warb. = *Burmannia disticha* L.
*Burmannia griffithii* Becc. = *Burmannia wallichii* (Miers) Hook.f.
*Burmannia herthae* G.M.Schulze = *Burmannia kalbreyeri* Oliv.
*Burmannia hunanensis* K.M.Liu & C.L.Long = *Burmannia championii* Thwaites
*Burmannia inhambanensis* Schltr. = *Burmannia madagascariensis* Mart.
*Burmannia japonica* Maxim. ex Matsum. = *Burmannia championii* Thwaites
*Burmannia javanica* Blume = *Burmannia coelestis* D.Don
*Burmannia le-testui* Schltr. = *Burmannia latialata* Pobég.
*Burmannia leucantha* Schltr. = *Burmannia longifolia* Becc.
*Burmannia liberica* Engl. = *Burmannia latialata* Pobég.
*Burmannia liukiuensis* Hayata = *Burmannia nepalensis* (Miers) Hook.f.
*Burmannia madagascariensis* Baker = *Burmannia madagascariensis* Mart.

*Burmannia malaccensis* Gand. = **Burmannia coelestis** D.Don

*Burmannia nana* Fukuy. & T.Suzuki = **Gymnosiphon aphyllus** Blume

*Burmannia novae-hiberniae* Schltr. = **Burmannia lutescens** Becc.

*Burmannia obscurata* Schltr. = **Burmannia latialata** Pobég.

*Burmannia paniculata* Willd. ex Schult. & Schult.f. = **Burmannia madagascariensis** Mart.

*Burmannia papillosa* Stapf = **Burmannia lutescens** Becc.

*Burmannia pingbienensis* H.Li = **Burmannia itoana** Makino

*Burmannia pusilla* var. *hongkongensis* Jonker = **Burmannia chinensis** Gand.

*Burmannia quadriflora* Willd. ex Schult. & Schult.f. = **Burmannia bicolor** Mart.

*Burmannia rigida* Gand. = **Burmannia chinensis** Gand.

*Burmannia selebica* Becc. = **Burmannia coelestis** D.Don

*Burmannia sellowiana* Seub. = **Burmannia alba** Mart.

*Burmannia sellowiana* var. *albiflora* Seub. = **Burmannia alba** Mart.

*Burmannia sellowiana* var. *violacea* Seub. = **Burmannia alba** Mart.

*Burmannia sumatrana* Miq. = **Burmannia disticha** L.

*Burmannia takeoi* Hayata = **Burmannia itoana** Makino

*Burmannia tridentata* Becc. = **Burmannia lutescens** Becc.

*Burmannia triflora* Roxb. = **Burmannia coelestis** D.Don

*Burmannia tuberosa* Becc. = **Burmannia championii** Thwaites

*Burmannia uniflora* Rottler ex Spreng. = **Burmannia coelestis** D.Don

*Burmannia urazii* Masam. = **Burmannia chinensis** Gand.

*Burmannia welwitschii* Schltr. = **Burmannia latialata** Pobég.

*Burmannia wercklei* Schltr. = **Burmannia kalbreyeri** Oliv.

## Campylosiphon

**Campylosiphon** Benth., Hooker's Icon. Pl. 14: t. 1384 (1882).
S. Trop. America. 82 FRG GUY SUR VEN 83 CLM 84 BZC BZN.
1 Species
*Dipterosiphon* Huber, Bol. Mus. Paraense Hist. Nat. Ethnogr. 2: 502 (1898).

**Campylosiphon purpurascens** Benth., Hooker's Icon. Pl. 14: t. 1384 (1882).
S. Trop. America. 82 FRG GUY SUR VEN 83 CLM 84 BZC BZN. Holopar. tuber geophyte.
*Dipterosiphon spelaeicola* Huber, Bol. Mus. Paraense Hist. Nat. Ethnogr. 2: 502 (1898).

**Synonyms:**
*Campylosiphon lycioideus* St.-Lag. = (Campanulaceae)

## Cryptonema

*Cryptonema* Turcz. = **Burmannia** L.
*Cryptonema malaccensis* Turcz. = **Burmannia coelestis** D.Don

## Cyananthus

*Cyananthus pusillus* (Miers) Miers = **Burmannia pusilla** (Miers) Thwaites

## Cyanotis

*Cyanotis nepalensis* (Miers) Miers = **Burmannia nepalensis** (Miers) Hook.f.

## Cymbocarpa

**Cymbocarpa** Miers, Proc. Linn. Soc. London 1: 61 (1840).
Trop. America. 80 COS PAN 81 CUB DOM HAI JAM PUE? 82 GUY VEN 83 CLM PER 84 BZL BZN BZS.
2 Species

**Cymbocarpa refracta** Miers, Proc. Linn. Soc. London 1: 62 (1840). *Gymnosiphon refractus* (Miers) Benth. & Hook.f., Gen. Pl. 3: 458 (1883).
Trop. America. 80 COS PAN 81 CUB DOM HAI JAM PUE? 82 VEN 83 CLM PER 84 BZL BZS. Holomycotrophic rhizome geophyte.
*Cymbocarpa urbanii* Goebel & Suess., Flora 117: 80 (1924).
*Cymbocarpa refracta* f. *albida* Cif., Atti Ist. Bot. Lab. Crittog. Univ. Pavia, V, 7: 27 (1946).

**Cymbocarpa saccata** Sandwith, Bull. Misc. Inform. Kew 1931: 60 (1931).
S. Trop. America. 82 GUY 83 PER 84 BZN. Holomycotrophic rhizome geophyte.

**Synonyms:**
*Cymbocarpa refracta* f. *albida* Cif. = **Cymbocarpa refracta** Miers
*Cymbocarpa urbanii* Goebel & Suess. = **Cymbocarpa refracta** Miers

## Desmogymnosiphon

**Desmogymnosiphon** Guinea, Ensayo Geobot. Guin. Continent. Espan.: 264 (1946).
WC. Trop. Africa. 23 EQG.
1 Species

**Desmogymnosiphon chimeicus** Guinea, Ensayo Geobot. Guin. Continent. Espan.: 264 (1946).
C. Equatorial Guinea. 23 EQG. Ther.

## Dictyostega

**Dictyostega** Miers, Proc. Linn. Soc. London 1: 61 (1840).
Trop. America. 79 MXG MXS MXT 80 BLZ COS GUA NIC PAN 81 TRT 82 FRG GUY SUR VEN 83 BOL CLM ECU PER 84 BZE BZL BZN BZS.
1 Species

**Dictyostega orobanchoides** (Hook.) Miers, Proc. Linn. Soc. London 1: 61 (1840).
Trop. America. 79 MXG MXS MXT 80 BLZ COS GUA NIC PAN 81 TRT 82 FRG GUY SUR VEN 83 BOL CLM ECU PER 84 BZE BZL BZN BZS. Holomycotroph.
*\*Apteria orobanchoides* Hook., Hooker's Icon. Pl. 3: t. 254 (1840).

subsp. *orobanchoides*
Trop. America. 79 MXG MXS MXT 80 BLZ COS GUA NIC PAN 81 TRT 82 GUY SUR VEN 83 BOL CLM ECU PER 84 BZE BZL BZN BZS. Holomycotroph.
*Dictyostega schomburgkii* Miers, Proc. Linn. Soc. London 1: 61 (1840).

subsp. *parviflora* (Benth.) Snelders & Maas, Acta Bot. Neerl. 30: 143 (1981).

Trinidad, Panama to S. Trop. America. 80 PAN 81 TRT 82 FRG GUY SUR VEN 83 BOL CLM ECU 84 BZN BZS. Holomycotroph.
  *Dictyostega schomburgkii* var. *parviflora* Benth., Hooker's J. Bot. Kew Gard. Misc. 7: 13 (1855).
  *Dictyostega orobanchoides* var. *parviflora* (Benth.) Jonker in A.A.Pulle, Fl. Suriname 1(1): 185 (1938).
  *Dictyostega campanulata* H.Karst., Linnaea 28: 422 (1857).
  *Gymnosiphon orobanchoides* Rusby, Bull. New York Bot. Gard. 6: 496 (1910). *Ptychomeria orobanchoides* (Rusby) Schltr., Repert. Spec. Nov. Regni Veg. 17: 257 (1921).

subsp. **purdieana** (Benth.) Snelders & Maas, Acta Bot. Neerl. 30: 143 (1981).
  S. Panama to Peru. 80 PAN 83 CLM ECU PER. Holomycotroph.
    * *Dictyostega purdieana* Benth., Hooker's J. Bot. Kew Gard. Misc. 7: 14 (1855).
    *Dictyostega pectinata* H.Karst., Linnaea 28: 422 (1857).

*Synonyms*:
*Dictyostega campanulata* H.Karst. = **Dictyostega orobanchoides** subsp. **parviflora** (Benth.) Snelders & Maas
*Dictyostega costata* Miers = **Miersiella umbellata** (Miers) Urb.
*Dictyostega longistyla* Benth. = **Gymnosiphon longistylus** (Benth.) Hutch.
*Dictyostega orobanchoides* var. *parviflora* (Benth.) Jonker = **Dictyostega orobanchoides** subsp. **parviflora** (Benth.) Snelders & Maas
*Dictyostega pectinata* H.Karst. = **Dictyostega orobanchoides** subsp. **purdieana** (Benth.) Snelders & Maas
*Dictyostega purdieana* Benth. = **Dictyostega orobanchoides** subsp. **purdieana** (Benth.) Snelders & Maas
*Dictyostega schomburgkii* Miers = **Dictyostega orobanchoides** (Hook.) Miers subsp. **orobanchoides**
*Dictyostega schomburgkii* var. *parviflora* Benth. = **Dictyostega orobanchoides** subsp. **parviflora** (Benth.) Snelders & Maas
*Dictyostega umbellata* Miers = **Miersiella umbellata** (Miers) Urb.
*Dictyostega usambarica* (Engl.) Engl. = **Gymnosiphon usambaricus** Engl.

## Dipterosiphon

*Dipterosiphon* Huber = **Campylosiphon** Benth.
*Dipterosiphon spelaeicola* Huber = **Campylosiphon purpurascens** Benth.

## Geomitra

*Geomitra* Becc., Malesia 1: 250 (1878).
  W. Malesia. 42 BOR MLY SUM.
  1 Species

*Geomitra clavigera* Becc., Malesia 1: 251 (1878).
  *Thismia clavigera* (Becc.) F.Muell., Pap. & Proc. Roy. Soc. Tasmania 1890: 235 (1891). *Sarcosiphon clavigerus* (Becc.) Schltr., Notizbl. Bot. Gart. Berlin-Dahlem 8: 39 (1921).
  W. Malesia. 42 BOR MLY SUM. Holomycotrophic rhizome geophyte.

*Synonyms*:
*Geomitra episcopalis* Becc. = **Thismia episcopalis** (Becc.) F.Muell.

## Glaziocharis

*Glaziocharis* Taub. ex Warm. = **Thismia** Griff.
*Glaziocharis abei* Akasawa = **Thismia abei** (Akasawa) Hatus.
*Glaziocharis macahensis* Taub. ex Warm. = **Thismia caudata** Maas & H.Maas

## Gonianthes

*Gonianthes* Blume = **Burmannia** L.
*Gonianthes candida* Blume = **Burmannia lutescens** Becc.
*Gonianthes nepalensis* Miers = **Burmannia nepalensis** (Miers) Hook.f.
*Gonianthes pusilla* Miers = **Burmannia pusilla** (Miers) Thwaites
*Gonianthes wallichii* Miers = **Burmannia wallichii** (Miers) Hook.f.

## Gonyanthes

*Gonyanthes* Neïes = **Burmannia** L.

## Gymnosiphon

*Gymnosiphon* Blume, Enum. Pl. Javae 1: 29 (1827).
  Trop. 22 GHA IVO LBR NGA SIE 23 CAF CMN GAB 25 KEN TAN 29 MDG 38 TAI 41 THA 42 BOR JAW LSI MLY SUL SUM 43 NWG 62 CRL 79 MXE MXG MXS MXT 80 BLZ COS GUA HON NIC PAN 81 CUB DOM HAI JAM LEE PUE TRT WIN 82 FRG GUY SUR VEN 83 BOL CLM ECU PER 84 BZC BZE BZL BZN BZS.
  26 Species
  *Benitzia* H.Karst., Linnaea 28: 420 (1857).
  *Ptychomeria* Benth., Hooker's J. Bot. Kew Gard. Misc. 7: 14 (1955).

*Gymnosiphon affinis* J.J.Sm., Nova Guinea 8: 194 (1909).
  New Guinea. 43 NWG. Holomycotrophic ther.
    *Gymnosiphon torricellensis* Schltr., Bot. Jahrb. Syst. 49: 101 (1912).

*Gymnosiphon aphyllus* Blume, Enum. Pl. Javae 1: 29 (1827).
  Taiwan (Lan Yü) to Malesia and Caroline Is. 38 TAI 41 THA 42 BOR JAW LSI MLY SUL SUM 43 NWG 62 CRL. Holomycotrophic ther.
    *Gymnosiphon borneensis* Becc., Malesia 1: 241 (1878).
    *Gymnosiphon pedicellatus* Schltr., Bot. Jahrb. Syst. 49: 105 (1912).
    *Burmannia nana* Fukuy. & T.Suzuki, J. Jap. Bot. 12: 415 (1936). *Gymnosiphon nanus* (Fukuy. & T.Suzuki) Tuyama, Iconogr. Pl. Asiae Orient. 3: 239 (1940).

*Gymnosiphon bekensis* Letouzey, Adansonia, n.s., 7: 170 (1967).
  WC. Trop. Africa. 23 CAF CMN GAB.

*Gymnosiphon brachycephalus* Snelders & Maas, Acta Bot. Neerl. 30: 142 (1981).
  Panama to N. South America and Ecuador. 80 PAN 82 GUY SUR VEN 83 CLM ECU. Holomycotrophic rhizome geophyte.

*Gymnosiphon breviflorus* Gleason, Bull. Torrey Bot. Club 56: 22 (1929). *Ptychomeria breviflora* (Gleason) Brade, Arch. Jard. Bot. Rio de Janeiro 7: 22 (1947).

Costa Rica to S. Trop. America. 80 COS PAN 82 FRG GUY SUR VEN 83 BOL CLM PER 84 BZN. Holomycotrophic rhizome geophyte.

***Gymnosiphon capitatus*** (Benth.) Urb., Symb. Antill. 3: 439 (1903).
Guyana to N. Brazil. 82 GUY 84 BZN. Holomycotrophic rhizome geophyte.
*\*Ptychomeria capitata* Benth., Hooker's J. Bot. Kew Gard. Misc. 7: 15 (1855).

***Gymnosiphon cymosus*** (Benth.) Benth. & Hook.f., Gen. Pl. 3: 458 (1883).
S. Trop. America. 82 SUR VEN 83 CLM PER 84 BZN. Holomycotrophic rhizome geophyte.
*\*Ptychomeria cymosa* Benth., Hooker's J. Bot. Kew Gard. Misc. 7: 15 (1855).

***Gymnosiphon danguyanus*** H.Perrier, Notul. Syst. (Paris) 5: 160 (1936).
W. Tanzania, Madagascar. 25 TAN 29 MDG. Holomycotrophic ther.

***Gymnosiphon divaricatus*** (Benth.) Benth. & Hook.f., Gen. Pl. 3: 458 (1883).
Trinidad, S. Mexico to S. Trop. America. 79 MXS MXT 80 BLZ COS GUA HON PAN 81 TRT 82 FRG GUY SUR VEN 83 CLM PER 84 BZC BZE BZL BZN BZS. Holomycotrophic rhizome geophyte.
*Ptychomeria cornuta* Benth., Hooker's J. Bot. Kew Gard. Misc. 7: 16 (1855). *Gymnosiphon cornutus* (Benth.) Benth. & Hook.f., Gen. Pl. 3: 458 (1883).
*\*Ptychomeria divaricata* Benth., Hooker's J. Bot. Kew Gard. Misc. 7: 16 (1855).
*Ptychomeria mutica* Benth., Hooker's J. Bot. Kew Gard. Misc. 7: 16 (1855). *Gymnosiphon muticus* (Benth.) Urb., Symb. Antill. 3: 438 (1903).
*Ptychomeria tenella* var. *minor* Benth., Hooker's J. Bot. Kew Gard. Misc. 7: 17 (1855).
*Benitzia poeppigiana* H.Karst., Linnaea 28: 421 (1857). *Ptychomeria poeppigiana* (H.Karst.) Schltr., Repert. Spec. Nov. Regni Veg. 17: 257 (1921).
*Gymnosiphon arcuatus* Urb., Symb. Antill. 3: 443 (1903). *Ptychomeria arcuata* (Urb.) Brade, Arch. Jard. Bot. Rio de Janeiro 7: 23 (1947).
*Ptychomeria mattogrossensis* Malme, Ark. Bot. 26A(9): 21 (1935). *Gymnosiphon mattogrossensis* (Malme) Jonker, Monogr. Burmann.: 192 (1938).
*Gymnosiphon tuerckheimii* Jonker, Monogr. Burmann.: 197 (1938).

***Gymnosiphon fimbriatus*** (Benth.) Urb., Symb. Antill. 3: 438 (1903).
S. Trop. America. 82 GUY VEN 83 CLM 84 BZN. Holomycotrophic rhizome geophyte.
*\*Ptychomeria fimbriata* Benth., Hooker's J. Bot. Kew Gard. Misc. 7: 14 (1855).

***Gymnosiphon guianensis*** Gleason, Bull. Torrey Bot. Club 56: 22 (1929). *Ptychomeria guianensis* (Gleason) Brade, Arch. Jard. Bot. Rio de Janeiro 7: 22 (1947).
Guianas. 82 GUY SUR. Holomycotrophic rhizome geophyte.

***Gymnosiphon longistylus*** (Benth.) Hutch. in J.Hutchinson & J.M.Dalziel, Fl. W. Trop. Afr. 2: 399 (1936).
W. & WC. Trop. Africa. 22 GHA IVO LBR NGA SIE 23 CMN GAB. Holomycotrophic ther.
*\*Dictyostega longistyla* Benth. in W.J.Hooker, Niger Fl.: 528 (1849).

***Gymnosiphon squamatus*** C.H.Wright, Bull. Misc. Inform. Kew 1897: 281 (1897). *Ptychomeria squamata* (C.H.Wright) Schltr., Repert. Spec. Nov. Regni Veg. 17: 258 (1921).

***Gymnosiphon minahassae*** Schltr., Bot. Jahrb. Syst. 49: 104 (1912).
N. Sulawesi. 42 SUL. Holomycotrophic ther.

***Gymnosiphon minutus*** Snelders & Maas, Acta Bot. Neerl. 30: 142 (1981).
Costa Rica to S. Trop. America. 80 COS 82 GUY VEN 83 CLM PER 84 BZN.

***Gymnosiphon neglectus*** Jonker, Monogr. Burmann.: 175 (1938).
Jawa. 42 JAW. Holomycotrophic ther.

***Gymnosiphon niveus*** (Griseb.) Urb., Symb. Antill. 3: 444 (1903).
Caribbean. 81 CUB DOM HAI JAM LEE PUE TRT WIN. Holomycotrophic hemicr.
*\*Ptychomeria nivea* Griseb., Cat. Pl. Cub.: 257 (1866).
*Gymnosiphon germainii* Urb., Symb. Antill. 3: 444 (1903). *Ptychomeria germainii* (Urb.) Stehlé, Bull. Soc. Bot. France 85: 514 (1938 publ. 1939).
*Gymnosiphon parviflorus* Urb., Symb. Antill. 3: 443 (1903). *Ptychomeria parviflora* (Urb.) Schltr., Repert. Spec. Nov. Regni Veg. 17: 257 (1921).
*Gymnosiphon portoricensis* Urb., Symb. Antill. 3: 445 (1903). *Ptychomeria portoricensis* (Urb.) Schltr., Repert. Spec. Nov. Regni Veg. 17: 257 (1921).
*Gymnosiphon fawcettii* Urb., Symb. Antill. 5: 294 (1907). *Ptychomeria fawcettii* (Urb.) Schltr., Repert. Spec. Nov. Regni Veg. 17: 257 (1921).
*Ptychomeria portoricensis* f. *roseocyanea* Cif., Atti Ist. Bot. Lab. Crittog. Univ. Pavia, V, 7: 27 (1946).

***Gymnosiphon okamotoi*** Tuyama, Iconogr. Pl. Asiae Orient. 3: 239 (1940).
Caroline Is. (Palau). 62 CRL.

***Gymnosiphon oliganthus*** Schltr., Bot. Jahrb. Syst. 49: 101 (1912).
NE. New Guinea. 43 NWG. Holomycotrophic ther.

***Gymnosiphon panamensis*** Jonker, Monogr. Burmann.: 199 (1938).
Mexico (Veracruz, Oaxaca, Chiapas) to C. America. 79 MXG MXS MXT 80 BLZ COS HON NIC PAN. Holomycotrophic rhizome geophyte.

***Gymnosiphon papuanus*** Becc., Malesia 1: 241 (1878).
Sulawesi to Caroline Is. (Palau). 42 SUL 43 NWG 62 CRL. Holomycotrophic ther.
*Gymnosiphon celebicus* Schltr., Bot. Jahrb. Syst. 49: 104 (1912).

***Gymnosiphon pauciflorus*** Schltr., Bot. Jahrb. Syst. 49: 102 (1912).
New Guinea (Kani Mts.). 43 NWG. Holomycotrophic ther.

***Gymnosiphon recurvatus*** Snelders & Maas, Acta Bot. Neerl. 30: 141 (1981).
Guyana. 82 GUY. Holomycotrophic rhizome geophyte.

***Gymnosiphon sphaerocarpus*** Urb., Symb. Antill. 3: 442 (1903). *Ptychomeria sphaerocarpa* (Urb.) Schltr., Repert. Spec. Nov. Regni Veg. 17: 257 (1921).
Caribbean. 81 CUB DOM JAM LEE PUE WIN. Holomycotrophic hemicr.

***Gymnosiphon suaveolens*** (H.Karst.) Urb., Symb. Antill. 3: 438 (1903).

Mexico to W. South America. 79 MXE MXG MXS
MXT 80 COS GUA HON NIC PAN 82 VEN 83
CLM ECU PER. Holomycotrophic rhizome
geophyte.
*Benitzia suaveolens H.Karst., Linnaea 28: 420
(1857). Ptychomeria suaveolens (H.Karst.) Schltr.,
Repert. Spec. Nov. Regni Veg. 17: 257 (1921).

**Gymnosiphon tenellus** (Benth.) Urb., Symb. Antill. 3:
438 (1903).
Jamaica, C. & S. Trop. America. 80 COS HON PAN
81 JAM 82 VEN 83 CLM BZL BZN.
Holomycotrophic rhizome geophyte.
*Ptychomeria tenella Benth., Hooker's J. Bot. Kew
Gard. Misc. 7: 17 (1855).
Gymnosiphon glaziovii Urb., Symb. Antill. 3: 438
(1903). Ptychomeria glaziovii (Urb.) Schltr., Repert.
Spec. Nov. Regni Veg. 17: 257 (1921).
Gymnosiphon pusillus Urb., Symb. Antill. 3: 438
(1903). Ptychomeria pusilla (Urb.) Schltr., Repert.
Spec. Nov. Regni Veg. 17: 257 (1921).
Gymnosiphon jamaicensis Urb., Symb. Antill. 5: 293
(1907). Ptychomeria jamaicensis (Urb.) Schltr.,
Repert. Spec. Nov. Regni Veg. 17: 257 (1921).

**Gymnosiphon usambaricus** Engl., Bot. Jahrb. Syst. 20:
138 (1894). Dictyostega usambarica (Engl.) Engl.,
Abh. Königl. Akad. Wiss. Berlin 39: 45 (1894).
Ptychomeria usambarica (Engl.) Schltr., Repert.
Spec. Nov. Regni Veg. 17: 258 (1921).
SE. Kenya to Tanzania. 25 KEN TAN.
Holomycotrophic ther.

*Synonyms:*
Gymnosiphon altsonii Gleason = **Hexapterella
gentianoides** Urb.
Gymnosiphon arcuatus Urb. = **Gymnosiphon
divaricatus** (Benth.) Benth. & Hook.f.
Gymnosiphon borneensis Becc. = **Gymnosiphon
aphyllus** Blume
Gymnosiphon celebicus Schltr. = **Gymnosiphon
papuanus** Becc.
Gymnosiphon congestus C.H.Wright = **Burmannia
congesta** (C.H.Wright) Jonker
Gymnosiphon cornutus (Benth.) Benth. & Hook.f. =
**Gymnosiphon divaricatus** (Benth.) Benth. & Hook.f.
Gymnosiphon fawcettii Urb. = **Gymnosiphon niveus**
(Griseb.) Urb.
Gymnosiphon germainii Urb. = **Gymnosiphon niveus**
(Griseb.) Urb.
Gymnosiphon glaziovii Urb. = **Gymnosiphon tenellus**
(Benth.) Urb.
Gymnosiphon jamaicensis Urb. = **Gymnosiphon tenellus**
(Benth.) Urb.
Gymnosiphon mattogrossensis (Malme) Jonker =
**Gymnosiphon divaricatus** (Benth.) Benth. & Hook.f.
Gymnosiphon muticus (Benth.) Urb. = **Gymnosiphon
divaricatus** (Benth.) Benth. & Hook.f.
Gymnosiphon nanus (Fukuy. & T.Suzuki) Tuyama =
**Gymnosiphon aphyllus** Blume
Gymnosiphon orobanchoides Rusby = **Dictyostega
orobanchoides** subsp. **parviflora** (Benth.) Snelders &
Maas
Gymnosiphon parviflorus Urb. = **Gymnosiphon niveus**
(Griseb.) Urb.
Gymnosiphon pedicellatus Schltr. = **Gymnosiphon
aphyllus** Blume
Gymnosiphon portoricensis Urb. = **Gymnosiphon niveus**
(Griseb.) Urb.
Gymnosiphon pusillus Urb. = **Gymnosiphon tenellus**
(Benth.) Urb.

Gymnosiphon refractus (Miers) Benth. & Hook.f. =
**Cymbocarpa refracta** Miers
Gymnosiphon squamatus C.H.Wright = **Gymnosiphon
longistylus** (Benth.) Hutch.
Gymnosiphon torricellensis Schltr. = **Gymnosiphon
affinis** J.J.Sm.
Gymnosiphon trinitatis Johow = **Marthella trinitatis**
(Johow) Urb.
Gymnosiphon tuerckheimii Jonker = **Gymnosiphon
divaricatus** (Benth.) Benth. & Hook.f.

## Haplothismia

**Haplothismia** Airy Shaw, Kew Bull. 7: 277 (1952).
S. India. 40 IND.
1 Species

**Haplothismia exannulata** Airy Shaw, Kew Bull. 7: 277
(1952).
S. India. 40 IND. Holomycotrophic hemicr.

## Hexapterella

**Hexapterella** Urb., Symb. Antill. 3: 451 (1903).
Colombia to Trinidad and N. Brazil. 81 TRT 82 FRG
GUY SUR VEN 83 CLM 84 BZN.
2 Species

**Hexapterella gentianoides** Urb., Symb. Antill. 3: 451
(1903).
Colombia to Trinidad and N. Brazil. 81 TRT 82 FRG
GUY SUR VEN 83 CLM 84 BZN. Holomycotrophic
rhizome geophyte.
Gymnosiphon altsonii Gleason, Bull. Torrey Bot. Club
56: 23 (1929).

**Hexapterella steyermarkii** Maas & H.Maas, Ann.
Missouri Bot. Gard. 76: 956 (1989).
S. Venezuela. 82 VEN. Holomycotrophic rhizome
geophyte.

## Maburnia

*Maburnia* Thouars = **Burmannia** L.

## Mamorea

*Mamorea* de la Sota = **Thismia** Griff.
*Mamorea singeri* de la Sota = **Thismia singeri** (de la
Sota) Maas & H.Maas

## Marthella

**Marthella** Urb., Symb. Antill. 3: 440 (1903).
Trinidad. 81 TRT.
1 Species

**Marthella trinitatis** (Johow) Urb., Symb. Antill. 3: 448
(1903).
N. Trinidad. 81 TRT. Holomycotroph.
*Gymnosiphon trinitatis Johow, Jahrb. Wiss. Bot. 20:
477 (1889).

## Miersiella

**Miersiella** Urb., Symb. Antill. 3: 439 (1903).
S. Trop. America. 82 FRG GUY SUR VEN 83 CLM
PER 84 BZE BZL BZS.
1 Species

**Miersiella umbellata** (Miers) Urb., Symb. Antill. 3: 439
(1903).
S. Trop. America. 82 FRG GUY SUR VEN 83 CLM
PER 84 BZE BZL BZS. Holomycotrophic rhizome
geophyte.

*Dictyostega costata* Miers, Proc. Linn. Soc. London 1: 61 (1840). *Miersiella costata* (Miers) Sandwith, Bull. Misc. Inform. Kew 1931: 59 (1931).
*\*Dictyostega umbellata* Miers, Proc. Linn. Soc. London 1: 61 (1840).
*Miersiella aristata* Sandwith, Bull. Misc. Inform. Kew 1931: 59 (1931).
*Miersiella kuhlmannii* Brade, Rodriguésia 10(20): 41 (1946).

**Synonyms:**
*Miersiella aristata* Sandwith = **Miersiella umbellata** (Miers) Urb.
*Miersiella costata* (Miers) Sandwith = **Miersiella umbellata** (Miers) Urb.
*Miersiella kuhlmannii* Brade = **Miersiella umbellata** (Miers) Urb.

## Myostoma

*Myostoma* Miers = **Thismia** Griff.
*Myostoma hyalinum* Miers = **Thismia hyalina** (Miers) Benth. & Hook.f. ex F.Muell.
*Myostoma janeirense* (Warm.) Schltr. = **Thismia janeirensis** Warm.

## Nemitis

*Nemitis* Raf. = **Apteria** Nutt.
*Nemitis setacea* (Nutt.) Raf. = **Apteria aphylla** (Nutt.) Barnhart ex Small

## Nephrocoelium

*Nephrocoelium* Turcz. = **Burmannia** L.
*Nephrocoelium malaccensis* (Turcz.) Turcz. = **Burmannia coelestis** D.Don

## Ophiomeris

*Ophiomeris* Miers = **Thismia** Griff.
*Ophiomeris espirito-santensis* (Brade) Brade = **Thismia espirito-santensis** Brade
*Ophiomeris iguassuensis* Miers = **Thismia iguassuensis** (Miers) Warm.
*Ophiomeris janeirensis* (Warm.) Brade = **Thismia janeirensis** Warm.
*Ophiomeris luetzelburgii* (Goebel & Suess.) Brade = **Thismia luetzelburgii** Goebel & Suess.
*Ophiomeris macahensis* Miers = **Thismia macahensis** (Miers) F.Muell.
*Ophiomeris panamensis* Standl. = **Thismia panamensis** (Standl.) Jonker

## Oxygyne

**Oxygyne** Schltr., Bot. Jahrb. Syst. 38: 140 (1906).
Cameroon, S. Japan to Nansei-shoto. 23 CMN 38 JAP NNS.
3 Species
*Saionia* Hatus., J. Geobot. 24: 2 (1976).

**Oxygyne hyodoi** C.Abe & Akasawa, J. Jap. Bot. 64: 161 (1989).
Japan (Shikoku: Ehime Pref.). 38 JAP. Holomycotrophic rhizome geophyte.

**Oxygyne shinzatoi** (Hatus.) C.Abe & Akasawa, J. Jap. Bot. 64: 163 (1989).
Nansei-shoto. 38 NNS. Holomycotrophic rhizome geophyte.
*\*Saionia shinzatoi* Hatus., J. Geobot. 24: 2 (1976).

**Oxygyne triandra** Schltr., Bot. Jahrb. Syst. 38: 140 (1906).
Cameroon. 23 CMN. Holomycotrophic ther.

## Ptychomeria

*Ptychomeria* Benth. = **Gymnosiphon** Blume
*Ptychomeria arcuata* (Urb.) Brade = **Gymnosiphon divaricatus** (Benth.) Benth. & Hook.f.
*Ptychomeria breviflora* (Gleason) Brade = **Gymnosiphon breviflorus** Gleason
*Ptychomeria capitata* Benth. = **Gymnosiphon capitatus** (Benth.) Urb.
*Ptychomeria cornuta* Benth. = **Gymnosiphon divaricatus** (Benth.) Benth. & Hook.f.
*Ptychomeria cymosa* Benth. = **Gymnosiphon cymosus** (Benth.) Benth. & Hook.f.
*Ptychomeria divaricata* Benth. = **Gymnosiphon divaricatus** (Benth.) Benth. & Hook.f.
*Ptychomeria fawcettii* (Urb.) Schltr. = **Gymnosiphon niveus** (Griseb.) Urb.
*Ptychomeria fimbriata* Benth. = **Gymnosiphon fimbriatus** (Benth.) Urb.
*Ptychomeria germainii* (Urb.) Stehlé = **Gymnosiphon niveus** (Griseb.) Urb.
*Ptychomeria glaziovii* (Urb.) Schltr. = **Gymnosiphon tenellus** (Benth.) Urb.
*Ptychomeria guianensis* (Gleason) Brade = **Gymnosiphon guianensis** Gleason
*Ptychomeria jamaicensis* (Urb.) Schltr. = **Gymnosiphon tenellus** (Benth.) Urb.
*Ptychomeria mattogrossensis* Malme = **Gymnosiphon divaricatus** (Benth.) Benth. & Hook.f.
*Ptychomeria mutica* Benth. = **Gymnosiphon divaricatus** (Benth.) Benth. & Hook.f.
*Ptychomeria nivea* Griseb. = **Gymnosiphon niveus** (Griseb.) Urb.
*Ptychomeria orobanchoides* (Rusby) Schltr. = **Dictyostega orobanchoides** subsp. **parviflora** (Benth.) Snelders & Maas
*Ptychomeria parviflora* (Urb.) Schltr. = **Gymnosiphon niveus** (Griseb.) Urb.
*Ptychomeria poeppigiana* (H.Karst.) Schltr. = **Gymnosiphon divaricatus** (Benth.) Benth. & Hook.f.
*Ptychomeria portoricensis* (Urb.) Schltr. = **Gymnosiphon niveus** (Griseb.) Urb.
*Ptychomeria portoricensis* f. *roseocyanea* Cif. = **Gymnosiphon niveus** (Griseb.) Urb.
*Ptychomeria pusilla* (Urb.) Schltr. = **Gymnosiphon tenellus** (Benth.) Urb.
*Ptychomeria sphaerocarpa* (Urb.) Schltr. = **Gymnosiphon sphaerocarpus** Urb.
*Ptychomeria squamata* (C.H.Wright) Schltr. = **Gymnosiphon longistylus** (Benth.) Hutch.
*Ptychomeria suaveolens* (H.Karst.) Schltr. = **Gymnosiphon suaveolens** (H.Karst.) Urb.
*Ptychomeria tenella* Benth. = **Gymnosiphon tenellus** (Benth.) Urb.
*Ptychomeria tenella* var. *minor* Benth. = **Gymnosiphon divaricatus** (Benth.) Benth. & Hook.f.
*Ptychomeria usambarica* (Engl.) Schltr. = **Gymnosiphon usambaricus** Engl.

## Rodwaya

*Rodwaya* F.Muell. = **Thismia** Griff.
*Rodwaya thismiacea* F.Muell. = **Thismia rodwayi** F.Muell.

## Saionia

*Saionia* Hatus. = *Oxygyne* Schltr.
*Saionia shinzatoi* Hatus. = *Oxygyne shinzatoi* (Hatus.)
C.Abe & Akasawa

## Sarcosiphon

*Sarcosiphon* Blume = *Thismia* Griff.
*Sarcosiphon americanus* (N.Pfeiff.) Schltr. = *Thismia americana* N.Pfeiff.
*Sarcosiphon clandestinum* Blume = *Thismia clandestina* (Blume) Miq.
*Sarcosiphon clavigerus* (Becc.) Schltr. = *Geomitra clavigera* Becc.
*Sarcosiphon croceus* (Becc.) Schltr. = *Thismia crocea* (Becc.) J.J.Sm.
*Sarcosiphon episcopalis* (Becc.) Schltr. = *Thismia episcopalis* (Becc.) F.Muell.
*Sarcosiphon hillii* (Cheeseman) Schltr. = *Thismia rodwayi* F.Muell.
*Sarcosiphon rodwayi* (F.Muell.) Schltr. = *Thismia rodwayi* F.Muell.
*Sarcosiphon versteegii* (J.J.Sm.) Schltr. = *Thismia crocea* (Becc.) J.J.Sm.

## Scaphiophora

**Scaphiophora** Schltr., Notizbl. Bot. Gart. Berlin-Dahlem 8: 39 (1921).
Philippines, New Guinea. 42 PHI 43 NWG.
2 Species

**Scaphiophora appendiculata** (Schltr.) Schltr., Notizbl. Bot. Gart. Berlin-Dahlem 8: 39 (1921).
NE. New Guinea. 43 NWG. Holomycotrophic rhizome geophyte.
*Thismia appendiculata* Schltr., Bot. Jahrb. Syst. 55: 202 (1918).

**Scaphiophora gigantea** Jonker, Monogr. Burmann.: 257 (1938).
Philippines (Luzon). 42 PHI. Holomycotrophic rhizome geophyte.

## Stemoptera

*Stemoptera* Miers = *Apteria* Nutt.
*Stemoptera lilacina* Miers = *Apteria aphylla* (Nutt.) Barnhart ex Small

## Tetraptera

*Tetraptera* Miers = *Burmannia* L.

## Thismia

**Thismia** Griff., Proc. Linn. Soc. London 1: 221 (1845).
Trop. Asia to Japan, E. & SE. Australia to New Zealand, NC. U.S.A., Costa Rica to S. Trop. America. 38 JAP TAI 40 SRL 41 MYA THA VIE 42 BOR JAW MLY SUM 43 NWG 50 NSW QLD TAS VIC 51 NZN 74 ILL† 80 COS PAN 82 FRG 83 BOL CLM ECU PER 84 BZL.
41 Species
*Ophiomeris* Miers, Proc. Linn. Soc. London 1: 328 (1847).
*Sarcosiphon* Blume, Mus. Bot. 1: 65 (1849).
*Tribrachys* Champ. ex Thwaites, Enum. Pl. Zeyl.: 325 (1864).
*Myostoma* Miers, Trans. Linn Soc. London 25: 474 (1866).
*Bagnisia* Becc., Malesia 1: 249 (1878).

*Rodwaya* F.Muell., Victoria Naturalist 7: 116 (1890), nom. inval.
*Triscyphus* Taub., Verh. Bot. Vereins Prov. Brandenburg 36: 66 (1895).
*Glaziocharis* Taub. ex Warm., Overs. Kongel. Danske Vidensk. Selsk. Forh. Medlemmers Arbeider 1901(6): 175 (1902).
*Triurocodon* Schltr., Notizbl. Bot. Gart. Berlin-Dahlem 8: 41 (1921).
*Mamorea* de la Sota, Darwiniana 12: 43 (1960).

**Thismia abei** (Akasawa) Hatus., J. Geobot. 24: 7 (1976).
S. Japan (Shikoku: Awa Prov.). 38 JAP. Holomycotrophic rhizome geophyte.
*Glaziocharis abei* Akasawa, J. Jap. Bot. 25: 193 (1950).

**Thismia alba** Holttum ex Jonker, in Fl. Males. 4: 23 (1948).
Pen. Malaysia (Pahang). 42 MLY. Holomycotrophic rhizome geophyte.

**Thismia americana** N.Pfeiff., Bot. Gaz. 57: 123 (1914).
*Sarcosiphon americanus* (N.Pfeiff.) Schltr., Notizbl. Bot. Gart. Berlin-Dahlem 8: 39 (1921).
NE. Illinois. 74 ILL†. Holomycotrophic rhizome geophyte.

**Thismia arachnites** Ridl., J. Straits Branch Roy. Asiat. Soc. 44: 197 (1905).
Pen. Malaysia. 42 MLY. Holomycotrophic rhizome geophyte.

**Thismia aseroe** Becc., Malesia 1: 252 (1878).
Malaya (Perak, Singapore). 42 MLY. Holomycotrophic rhizome geophyte.

**Thismia bifida** M.Hotta, Acta Phytotax. Geobot. 22: 161 (1967).
Borneo (Sarawak). 42 BOR. Holomycotrophic rhizome geophyte.

**Thismia brunonis** Griff., Proc. Linn. Soc. London 1: 221 (1845).
Myanmar. 41 MYA. Holomycotrophic rhizome geophyte.
*Thismia brunoniana* Griff., Trans. Linn. Soc. London 19: 341 (1845).

**Thismia caudata** Maas & H.Maas, Fl. Neotrop. Monogr. 42: 162 (1986).
Brazil (Rio de Janeiro). 84 BZL. Holomycotrophic rhizome geophyte.
*Glaziocharis macahensis* Taub. ex Warm., Overs. Kongel. Danske Vidensk. Selsk. Forh. Medlemmers Arbeider 1901(6): 175 (1902). *Thismia macahensis* (Taub. ex Warm.) Hatus., J. Geobot. 24: 7 (1976), nom. illeg.

**Thismia chrysops** Ridl., Ann. Bot. (Oxford) 9: 323 (1895).
Pen. Malaysia (Mt. Ophir). 42 MLY. Holomycotrophic rhizome geophyte.

**Thismia clandestina** (Blume) Miq., Fl. Ned. Ind. 3: 616 (1859).
W. Jawa. 42 JAW. Holomycotrophic rhizome geophyte.
*Sarcosiphon clandestinum* Blume, Mus. Bot. 1: 65 (1849).

**Thismia clavarioides** K.R.Thiele, Telopea 9: 766 (2002).
CE. New South Wales. 50 NSW. Holomycotrophic rhizome geophyte.

**Thismia crocea** (Becc.) J.J.Sm., Nova Guinea 8(1): 193 (1909).
W. New Guinea. 43 NWG. Holomycotrophic rhizome geophyte.
*\*Bagnisia crocea* Becc., Malesia 1: 249 (1878).
*Sarcosiphon croceus* (Becc.) Schltr., Notizbl. Bot. Gart. Berlin-Dahlem 8: 38 (1921).
*Thismia versteegii* J.J.Sm., Nova Guinea 8: 193 (1909).
*Sarcosiphon versteegii* (J.J.Sm.) Schltr., Notizbl. Bot. Gart. Berlin-Dahlem 8: 38 (1921).

**Thismia episcopalis** (Becc.) F.Muell., Pap. & Proc. Roy. Soc. Tasmania 1890: 235 (1891).
N. & NW. Borneo. 42 BOR. Holomycotrophic rhizome geophyte.
*\*Geomitra episcopalis* Becc., Malesia 1: 250 (1878).
*Bagnisia episcopalis* (Becc.) Engl. in H.G.A.Engler & K.A.E.Prantl, Nat. Pflanzenfam. 2(6): 48 (1888).
*Sarcosiphon episcopalis* (Becc.) Schltr., Notizbl. Bot. Gart. Berlin-Dahlem 8: 38 (1921).

**Thismia espirito-santensis** Brade, Revista Brasil. Biol. 7: 286 (1947). *Ophiomeris espirito-santensis* (Brade) Brade, Arch. Jard. Bot. Rio de Janeiro 7: 28 (1947).
Brazil (Espírito Santo). 84 BZL. Holomycotrophic tuber geophyte.

**Thismia fumida** Ridl., J. Straits Branch Roy. Asiat. Soc. 22: 338 (1890).
Pen. Malaysia (Selangor, Singapore). 42 MLY. Holomycotrophic rhizome geophyte.

**Thismia fungiformis** (Taub. ex Warm.) Maas & H.Maas, Fl. Neotrop. Monogr. 42: 165 (1986).
Brazil (Rio de Janeiro). 84 BZL. Holomycotrophic rhizome geophyte.
*\*Triscyphus fungiformis* Taub. ex Warm., Overs. Kongel. Danske Vidensk. Selsk. Forh. Medlemmers Arbeider 1901(6): 178 (1902).

**Thismia gardneriana** Hook.f. ex Thwaites, Enum. Pl. Zeyl.: 325 (1864).
Sri Lanka (Ratnapura Distr.). 40 SRL. Holomycotrophic rhizome geophyte.
*T. ribrachys gardneriana* Champ. ex Thwaites, Enum. Pl. Zeyl.: 325 (1864).

**Thismia glaziovii** Poulsen, Rev. Gén. Bot. 1: 549 (1889).
*Triurocodon glaziovii* (Poulsen) Schltr., Notizbl. Bot. Gart. Berlin-Dahlem 8: 41 (1921).
Brazil (Rio de Janeiro). 84 BZL. Holomycotrophic tuber geophyte.
*Thismia itatiaiensis* Brade, Arq. Serv. Florest. 2(1): 47 (1943). *Triurocodon itatiaiensis* (Brade) Brade, Arch. Jard. Bot. Rio de Janeiro 7: 27 (1947).

**Thismia goodii** Kiew, Gard. Bull. Singapore 51: 179 (1999).
Borneo (Sabah). 42 BOR. Holomycotrophic tuber geophyte.

**Thismia grandiflora** Ridl., Ann. Bot. (Oxford) 9: 324 (1895).
Pen. Malaysia (Johor). 42 MLY. Holomycotrophic rhizome geophyte.

**Thismia hyalina** (Miers) Benth. & Hook.f. ex F.Muell., Pap. & Proc. Roy. Soc. Tasmania 1890: 234 (1891).
N. Peru, SE. Brazil. 83 PER 84 BZL. Holomycotrophic tuber geophyte.
*\*Myostoma hyalinum* Miers, Trans. Linn. Soc. London 25: 474 (1866).

**Thismia iguassuensis** (Miers) Warm., Overs. Kongel. Danske Vidensk. Selsk. Forh. Medlemmers Arbeider 1901(6): 182 (1902).
Brazil (Rio de Janeiro). 84 BZL.
*\*Ophiomeris iguassuensis* Miers, Proc. Linn. Soc. London 1: 329 (1847).

**Thismia janeirensis** Warm., Overs. Kongel. Danske Vidensk. Selsk. Forh. Medlemmers Arbeider 1901: 183 (1902). *Myostoma janeirense* (Warm.) Schltr., Notizbl. Bot. Gart. Berlin-Dahlem 8: 41 (1921).
*Ophiomeris janeirensis* (Warm.) Brade, Arch. Jard. Bot. Rio de Janeiro 7: 28 (1947).
Brazil (S. Minas Gerais, Rio de Janeiro). 84 BZL. Holomycotrophic tuber geophyte.

**Thismia javanica** J.J.Sm., Ann. Jard. Bot. Buitenzorg 23: 32 (1910).
S. Indo-China to W. Malesia. 41 THA VIE 42 JAW SUM. Holomycotrophic rhizome geophyte.

**Thismia labiata** J.J.Sm., Bull. Jard. Bot. Buitenzorg, III, 9: 220 (1927).
E. Sumatera. 42 SUM. Holomycotrophic rhizome geophyte.

**Thismia lauriana** Jarvie, Blumea 41: 259 (1996).
Borneo (Kalimantan). 42 BOR. Holomycotrophic rhizome geophyte.

**Thismia luetzelburgii** Goebel & Suess., Flora 117: 56 (1924). *Ophiomeris luetzelburgii* (Goebel & Suess.) Brade, Arch. Jard. Bot. Rio de Janeiro 7: 28 (1947).
Costa Rica to Panama, Brazil (Espírito Santo). 80 COS PAN 84 BZL. Holomycotrophic tuber geophyte.

**Thismia macahensis** (Miers) F.Muell., Pap. & Proc. Roy. Soc. Tasmania 1890: 232 (1891).
Brazil (Rio de Janeiro). 84 BZL. Holomycotrophic tuber geophyte.
*\*Ophiomeris macahensis* Miers, Proc. Linn. Soc. London 1: 329 (1847).

**Thismia melanomitra** Maas & H.Maas, Opera Bot. 92: 141 (1987).
Ecuador. 83 ECU. Holomycotrophic rhizome geophyte.

**Thismia mirabilis** K.Larsen, Dansk Bot. Ark. 23: 171 (1965).
E. & SE. Thailand. 41 THA. Holomycotrophic rhizome geophyte.

**Thismia mullerensis** Tsukaya & H.Okada, Acta Phytotax. Geobot. 56: 129 (2005).
Borneo. 42 BOR. Holomycotrophic rhizome geophyte.

**Thismia neptunis** Becc., Malesia 1: 251 (1878).
Borneo (Sarawak: Mt. Matang). 42 BOR. Holomycotrophic rhizome geophyte.

**Thismia ophiuris** Becc., Malesia 1: 252 (1878).
N. & NW. Borneo. 42 BOR. Holomycotrophic rhizome geophyte.

**Thismia panamensis** (Standl.) Jonker, Monogr. Burmann.: 234 (1938).
Costa Rica to NE. Peru. 80 COS PAN 83 CLM ECU PER. Holomycotrophic tuber geophyte.
*\*Ophiomeris panamensis* Standl., J. Wash. Acad. Sci. 17: 163 (1927).

***Thismia racemosa*** Ridl., J. Straits Branch Roy. Asiat. Soc. 68: 13 (1915).
Pen. Malaysia (Pahang). 42 MLY. Holomycotrophic rhizome geophyte.

***Thismia rodwayi*** F.Muell., Victoria Naturalist 7: 115 (1890). *Rodwaya thismiacea* F.Muell., Victoria Naturalist 7: 116 (1890), nom. inval. *Bagnisia rodwayi* (F.Muell.) F.Muell., Pap. & Proc. Roy. Soc. Tasmania 1896: 232 (1891). *Sarcosiphon rodwayi* (F.Muell.) Schltr., Notizbl. Bot. Gart. Berlin-Dahlem 8: 39 (1921).
SE. Australia, New Zealand North I. 50 NSW TAS VIC 51 NZN. Holomycotrophic rhizome geophyte.
*Bagnisia hillii* Cheeseman, Bull. Misc. Inform. Kew 1908: 420 (1908). *Sarcosiphon hillii* (Cheeseman) Schltr., Notizbl. Bot. Gart. Berlin-Dahlem 8: 39 (1921).

***Thismia saulensis*** H.Maas & Maas, Brittonia 39: 376 (1987).
French Guiana. 82 FRG. Holomycotrophic rhizome geophyte.

***Thismia singeri*** (de la Sota) Maas & H.Maas, Fl. Neotrop. Monogr. 42: 166 (1986).
Bolivia (Beni). 83 BOL. Holomycotrophic rhizome geophyte.
*\*Mamorea singeri* de la Sota, Darwiniana 12: 45 (1960).

***Thismia taiwanensis*** Sheng Z.Yang, R.M.K.Saunders & C.J.Hsu, Syst. Bot. 27: 485 (2002).
SC. Taiwan. 38 TAI. Holomycotrophic rhizome geophyte.

***Thismia tuberculata*** Hatus., J. Geobot. 24: 4 (1976).
Japan. 38 JAP. Holomycotrophic rhizome geophyte.

***Thismia yorkensis*** Cribb, Queensland Naturalist 33: 51 (1995).
N. Queensland. 50 QLD. Holomycotrophic rhizome geophyte.

**Synonyms:**
*Thismia appendiculata* Schltr. = **Scaphiophora appendiculata** (Schltr.) Schltr.
*Thismia brunoniana* Griff. = **Thismia brunonis** Griff.
*Thismia clavigera* (Becc.) F.Muell. = **Geomitra clavigera** Becc.
*Thismia itatiaiensis* Brade = **Thismia glaziovii** Poulsen
*Thismia macahensis* (Taub. ex Warm.) Hatus. = **Thismia caudata** Maas & H.Maas
*Thismia pachyantha* (Schltr.) Engl. = **Afrothismia pachyantha** Schltr.
*Thismia versteegii* J.J.Sm. = **Thismia crocea** (Becc.) J.J.Sm.
*Thismia winkleri* Engl. = **Afrothismia winkleri** (Engl.) Schltr.

## Tiputinia

***Tiputinia*** P.E.Berry & C.L.Woodw., Taxon 56: 158 (2007). Ecuador. 83 ECU.
1 species.

***Tiputinia foetida*** P.E.Berry & C.L.Woodw., Taxon 56: 158 (2007).
E. Ecuador. 83 ECU. Holoparasitic rhizome geophyte.

## Tribrachys

*Tribrachys* Champ. ex Thwaites = **Thismia** Griff.
*Tribrachys gardneriana* Champ. ex Thwaites = **Thismia gardneriana** Hook.f. ex Thwaites

## Tripteranthus

*Tripteranthus* Wall. ex Miers = **Burmannia** L.

## Tripterella

*Tripterella* Michx. = **Burmannia** L.
*Tripterella alba* (Mart.) Schult. = **Burmannia alba** Mart.
*Tripterella bicolor* (Mart.) Schult. = **Burmannia bicolor** Mart.
*Tripterella biflora* (L.) Schult. = **Burmannia biflora** L.
*Tripterella caerulea* Muhl. ex Nutt. = **Burmannia biflora** L.
*Tripterella capitata* (Walter ex J.F.Gmel.) Michx. = **Burmannia capitata** (Walter ex J.F.Gmel.) Mart.
*Tripterella dasyantha* (Mart.) Schult. = **Burmannia dasyantha** Mart.
*Tripterella flava* (Mart.) Schult. = **Burmannia flava** Mart.

## Triscyphus

*Triscyphus* Taub. = **Thismia** Griff.
*Triscyphus fungiformis* Taub. ex Warm. = **Thismia fungiformis** (Taub. ex Warm.) Maas & H.Maas

## Triurocodon

*Triurocodon* Schltr. = **Thismia** Griff.
*Triurocodon glaziovii* (Poulsen) Schltr. = **Thismia glaziovii** Poulsen
*Triurocodon itatiaiensis* (Brade) Brade = **Thismia glaziovii** Poulsen

## Vogelia

*Vogelia* J.F.Gmel. = **Burmannia** L.
*Vogelia capitata* Walter ex J.F.Gmel. = **Burmannia capitata** (Walter ex J.F.Gmel.) Mart.

# Dioscoreaceae

## Ataccia

*Ataccia* C.Presl = *Tacca* J.R.Forst. & G.Forst.
*Ataccia aspera* (Roxb.) Kunth = *Tacca integrifolia* Ker Gawl.
*Ataccia cristata* (Jack) Kunth = *Tacca integrifolia* Ker Gawl.
*Ataccia integrifolia* (Ker Gawl.) C.Presl = *Tacca integrifolia* Ker Gawl.
*Ataccia laevis* (Roxb.) Kunth = *Tacca integrifolia* Ker Gawl.
*Ataccia lancifolia* (Zoll. & Moritzi) Kunth = *Tacca integrifolia* Ker Gawl.

## Avetra

*Avetra* H.Perrier = *Trichopus* Gaertn.
*Avetra sempervirens* H.Perrier = *Trichopus sempervirens* (H.Perrier) Caddick & Wilkin

## Borderea

*Borderea* Miégev. = *Dioscorea* Plum. ex L.
*Borderea chouardii* (Gaussen) Gaussen & Heslot = *Dioscorea chouardii* Gaussen
*Borderea humilis* (Bertero ex Colla) Pax = *Dioscorea humilis* Bertero ex Colla
*Borderea pyrenaica* Miégev. = *Dioscorea pyrenaica* Bubani & Bordère ex Gren.

## Botryosicyos

*Botryosicyos* Hochst. = *Dioscorea* Plum. ex L.
*Botryosicyos pentaphyllus* (L.) Hochst. = *Dioscorea pentaphylla* L.

## Chaitaea

*Chaitaea* Sol. ex Seem. = *Tacca* J.R.Forst. & G.Forst.
*Chaitaea tacca* Sol. ex Seem. = *Tacca leontopetaloides* (L.) Kuntze

## Dioscorea

**Dioscorea** Plum. ex L., Sp. Pl.: 1032 (1753).
Cosmopolitan. 10 GRB ire 11 AUT BGM GER HUN SWI 12 BAL COR FRA POR SAR SPA 13 ALB BUL GRC ITA KRI ROM SIC TUE YUG 14 KRY 20 LBY MOR TUN 21 AZO CNY MDR 22 BEN BKN GAM GHA GNB GUI IVO LBR MLI NGA NGR SEN SIE TOG 23 BUR CAB CAF CMN CON EQG GAB GGI RWA ZAI 24 CHA ERI ETH SOC SUD 25 KEN TAN UGA 26 ANG MLW MOZ ZAM ZIM 27 BOT CPP CPV NAM NAT OFS SWZ TVL 29 ALD COM MDG sey 31 AMU KHA KUR PRM 33 NCS TCS 34 CYP EAI IRN IRQ LBS PAL TUR 36 CHC CHH CHI CHM CHN CHQ CHS CHT 38 JAP KOR kzn NNS oga TAI 40 ASS BAN EHM IND NEP PAK SRL WHM 41 AND CBD LAO MYA NCB THA VIE 42 BOR JAW LSI MLY MOL PHI SUL SUM 43 BIS NWG 50 NSW NTA QLD WAU (51) nzn 60 fij nue sam ton VAN (61) mrq pit sci 62 CRL mrn (63) haw 72 ONT 74 ILL IOW KAN MIN MSO NEB OKL WIS 75 CNT INI MAS MIC NWJ NWY OHI PEN RHO VER WVA 77 TEX 78 ALA ARK DEL fka FLA GEO KTY LOU MRY MSI NCA SCA TEN VRG WDC 79 MXC MXE MXG MXN MXS MXT 80 BLZ COS ELS GUA HON NIC PAN 81 CUB DOM HAI JAM LEE PUE TRT WIN 82 FRG GUY SUR VEN 83 BOL CLM ECU PER 84 BZC BZE BZL BZN BZS 85 AGE AGS AGW CLC CLN CLS PAR URU.
608 Species
*Tamus* L., Sp. Pl.: 1028 (1753).
*Ricophora* Mill., Gard. Dict. Abr. ed. 4: s.p. (1754).
*Tamnus* Mill., Gard. Dict. Abr. ed. 4: s.p. (1754).
*Oncus* Lour., Fl. Cochinch.: 194 (1790).
*Ubium* J.F.Gmel., Syst. Nat. 2: 839 (1791).
*Oncorhiza* Pers., Syn. Pl. 1: 374 (1805).
*Testudinaria* Salisb. ex Burch., Trav. S. Africa 2: 147 (1824).
*Rhizemys* Raf., Fl. Tellur. 4: 26 (1838).
*Botryosicyos* Hochst., Flora 27(Beil.): 3 (1844).
*Helmia* Kunth, Enum. Pl. 5: 414 (1850).
*Sismondaea* Delponte, Mem. Reale Accad. Sci. Torino, II, 14: 394 (1854).
*Epipetrum* Phil., Linnaea 33: 253 (1865).
*Borderea* Miégev., Bull. Soc. Bot. France 13: 374 (1866).
*Elephantodon* Salisb., Gen. Pl.: 12 (1866).
*Hamatris* Salisb., Gen. Pl.: 11 (1866).
*Merione* Salisb., Gen. Pl.: 12 (1866).
*Polynome* Salisb., Gen. Pl.: 12 (1866).
*Strophis* Salisb., Gen. Pl.: 12 (1866).
*Higinbothamia* Uline, Publ. Field Columbian Mus., Bot. Ser. 1: 414 (1899).
*Nanarepenta* Matuda, Anales Inst. Biol. Univ. Nac. México 32: 143 (1962).
*Hyperocarpa* (Uline) G.M.Barroso, E.F.Guim. & Sucre, Sellowia 25: 19 (1974).

**Dioscorea abysmophila** Maguire & Steyerm., Mem. New York Bot. Gard. 51: 105 (1989).
Venezuela (Amazonas). 82 VEN. Cl. tuber geophyte.

**Dioscorea abyssinica** Hochst. ex Kunth, Enum. Pl. 5: 387 (1850).
W. Trop. Africa to Eritrea. 22 BEN GHA IVO LBR MLI NGA SEN 24 ERI ETH SUD. Cl. tuber geophyte.

**Dioscorea acanthogene** Rusby, Bull. New York Bot. Gard. 6: 492 (1910).
W. South America to Paraguay. 83 BOL CLM PER 84 BZC 85 PAR. Cl. tuber geophyte.
*Dioscorea multiflora* var. *gouanioides* Chodat & Hassl., Bull. Herb. Boissier, II, 3: 1111 (1903).
*Dioscorea gouanioides* (Chodat & Hassl.) R.Knuth, Notizbl. Bot. Gart. Berlin-Dahlem 7: 192 (1917).
*Dioscorea sulcata* R.Knuth, Notizbl. Bot. Gart. Berlin-Dahlem 7: 192 (1917), nom. illeg.
*Dioscorea corumbensis* R.Knuth in H.G.A.Engler (ed.), Pflanzenr., IV, 43: 241 (1924).
*Dioscorea guanaiensis* R.Knuth in H.G.A.Engler (ed.), Pflanzenr., IV, 43: 78 (1924).
*Dioscorea paraguayensis* R.Knuth in H.G.A.Engler (ed.), Pflanzenr., IV, 43: 78 (1924).

*Dioscorea narinensis* R.Knuth, Repert. Spec. Nov. Regni Veg. 28: 82 (1930).

*Dioscorea pozucoensis* R.Knuth, Repert. Spec. Nov. Regni Veg. 28: 86 (1930).

*Dioscorea apurimacensis* R.Knuth, Repert. Spec. Nov. Regni Veg. 29: 94 (1931).

**Dioscorea acerifolia** Phil., Anales Univ. Chile 93: 15 (1896).
SE. Chile. 85 CLS. Tuber geophyte.

**Dioscorea acuminata** Baker, J. Linn. Soc., Bot. 21: 449 (1885).
W. Madagascar. 29 MDG. Cl. tuber geophyte.
*Dioscorea bararum* H.Perrier, Mém. Soc. Linn. Normandie, Bot. 1(1): 33 (1928).

**Dioscorea adenantha** Uline in H.G.A.Engler & K.A.E.Prantl, Nat. Pflanzenfam., Nachtr. 1: 86 (1897).
SE. Brazil. 84 BZL. Cl. tuber geophyte.

**Dioscorea aesculifolia** R.Knuth in H.G.A.Engler (ed.), Pflanzenr., IV, 43: 170 (1924).
Brazil (?). 84+. Cl. tuber geophyte.

**Dioscorea aguilarii** Standl. & Steyerm., Publ. Field Mus. Nat. Hist., Bot. Ser. 22: 133 (1940).
Guatemala. 80 GUA. Cl. tuber geophyte.

**Dioscorea alata** L., Sp. Pl.: 1033 (1753). *Polynome alata* (L.) Salisb., Gen. Pl.: 12 (1866).
Trop. & Subtrop. Asia, cultivated elsewere. (22) ben tog (23) cmn gab rwa zai (25) tan (26) mlw moz zam (29) com mdg sey (36) chc chs (38) kzn NNS TAI 40 ASS EHM NEP 41 MYA THA VIE 42 BOR JAW LSI MLY PHI SUL SUM 43 BIS NWG (50) nta qld (60) fij nue sam ton (61) pit sci (62) crl mrn (78) fka geo (79) (80) blz cos (81) jam lee pue trt win (82) frg guy sur ven (83) per (84) bzc. Cl. tuber geophyte.
*Dioscorea eburina* Lour., Fl. Cochinch.: 625 (1790).
*Dioscorea eburnea* Lour., Fl. Cochinch.: 767 (1790). *Elephantodon eburnea* (Lour.) Salisb., Gen. Pl.: 12 (1866).
*Dioscorea atropurpurea* Roxb., Fl. Ind. ed. 1832, 3: 800 (1832).
*Dioscorea globosa* Roxb., Fl. Ind. ed. 1832, 3: 797 (1832). *Dioscorea alata* var. *globosa* (Roxb.) Prain, Bengal Pl. 2: 1067 (1903).
*Dioscorea purpurea* Roxb., Fl. Ind. ed. 1832, 3: 799 (1832).
*Dioscorea rubella* Roxb., Fl. Ind. ed. 1832, 3: 798 (1832).
*Dioscorea vulgaris* Miq., Fl. Ned. Ind. 3: 572 (1857).
*Dioscorea colocasiifolia* Pax, Bot. Jahrb. Syst. 15: 145 (1892).
*Dioscorea javanica* Queva, Mém. Soc. Natl. Sci. Agric. Arts Lille, IV, 20: 372 (1894).
*Dioscorea sapinii* De Wild., Ann. Mus. Congo Belge, Bot., V, 3: 368 (1912).
*Dioscorea alata* var. *tarri* Prain & Burkill, J. Proc. Asiat. Soc. Bengal 10: 39 (1914).
*Dioscorea alata* var. *vera* Prain & Burkill, J. Proc. Asiat. Soc. Bengal 10: 39 (1914).

**Dioscorea alatipes** Burkill & H.Perrier, Notul. Syst. (Paris) 14: 136 (1951).
SW. Madagascar. 29 MDG. Cl. tuber geophyte.

**Dioscorea althaeoides** R.Knuth in H.G.A.Engler (ed.), Pflanzenr., IV, 43: 180 (1924).
SC. China to Thailand. 36 CHC CHT 41 THA. Cl. rhizome geophyte.

*Dioscorea platanifolia* Prain & Burkill, Bull. Misc. Inform. Kew 1925: 60 (1925).

**Dioscorea altissima** Lam., Encycl. 3: 231 (1789).
Trop. America. 80 PAN 81 CUB DOM HAI LEE PUE TRT WIN 82 GUY SUR VEN 83 BOL CLM PER 84 BZC BZE BZL BZS. Cl. tuber geophyte.
*Dioscorea chondrocarpa* Griseb. in C.F.P.von Martius & auct. suc. (eds.), Fl. Bras. 3(1): 34 (1842).
*Dioscorea samydea* Griseb. in C.F.P.von Martius & auct. suc. (eds.), Fl. Bras. 3(1): 33 (1842).
*Dioscorea poeppigii* Kunth, Enum.. Pl. 5: 365 (1850). *Dioscorea samydea* var. *poeppigii* (Kunth) Ayala, Diosc. Peru: 47 (1998).
*Dioscorea riparia* Kunth & R.H.Schomb. in K.S.Kunth, Enum. Pl. 5: 364 (1850).
*Dioscorea nitida* R.Knuth, Notizbl. Bot. Gart. Berlin-Dahlem 7: 191 (1917).
*Dioscorea rajanioides* Uline ex R.Knuth, Notizbl. Bot. Gart. Berlin-Dahlem 7: 186 (1917).
*Dioscorea samydea* var. *corcovadensis* Uline ex R.Knuth, Notizbl. Bot. Gart. Berlin-Dahlem 7: 201 (1917).
*Dioscorea hoehneana* R.Knuth, Notizbl. Bot. Gart. Berlin-Dahlem 7: 538 (1921).
*Dioscorea calcarea* R.Knuth in H.G.A.Engler (ed.), Pflanzenr., IV, 43: 84 (1924).
*Dioscorea maranonensis* R.Knuth, Repert. Spec. Nov. Regni Veg. 22: 344 (1926).
*Dioscorea balsapuertensis* R.Knuth, Repert. Spec. Nov. Regni Veg. 38: 117 (1935).
*Dioscorea revillae* Ayala, Ann. Missouri Bot. Gard. 68: 125 (1981).

**Dioscorea amaranthoides** C.Presl, Reliq. Haenk. 1: 134 (1827).
Peru to NE. Argentina. 83 BOL PER 84 BZC BZL 85 AGE PAR. Cl. tuber geophyte.
*Dioscorea silvestris* Vell., Fl. Flumin. 10: t. 118 (1831).
*Dioscorea crumenigera* Mart. ex Griseb. in C.F.P.von Martius & auct. suc. (eds.), Fl. Bras. 3(1): 36 (1842). *Dioscorea amaranthoides* var. *crumenigera* (Mart. ex Griseb.) Uline ex R.Knuth, Notizbl. Bot. Gart. Berlin-Dahlem 7: 215 (1917).
*Dioscorea eldorado* Linden & André, Ill. Hort. 18: 52 (1871).
*Dioscorea multicolor* Linden & André, Ill. Hort. 18: 52 (1871).
*Dioscorea multicolor* var. *chrysophylla* Linden & André, Ill. Hort. 18: 52 (1871).
*Dioscorea multicolor* var. *melanoleuca* Linden & André, Ill. Hort. 18: 52 (1871).
*Dioscorea multicolor* var. *metallica* Linden & André, Ill. Hort. 18: 53 (1871).
*Dioscorea multicolor* var. *sagittaria* Linden & André, Ill. Hort. 18: 52 (1871).
*Dioscorea prismatica* Linden & André, Ill. Hort. 18: 52 (1871).
*Dioscorea apaensis* Chodat & Hassl., Bull. Herb. Boissier, II, 3: 1112 (1903).
*Dioscorea amaranthoides* var. *denudata* Uline ex R.Knuth, Notizbl. Bot. Gart. Berlin-Dahlem 7: 215 (1917).
*Dioscorea amaranthoides* var. *elegantula* Uline ex R.Knuth, Notizbl. Bot. Gart. Berlin-Dahlem 7: 215 (1917).
*Dioscorea amaranthoides* var. *glauca* Uline ex R.Knuth, Notizbl. Bot. Gart. Berlin-Dahlem 7: 215 (1917).
*Dioscorea amaranthoides* var. *metallica* Harms ex

R.Knuth, Notizbl. Bot. Gart. Berlin-Dahlem 7: 216 (1917).

*Dioscorea amaranthoides* var. *paniculata* R.Knuth, Notizbl. Bot. Gart. Berlin-Dahlem 7: 216 (1917).

**Dioscorea amazonum** Mart. ex Griseb. in C.F.P.von Martius & auct. suc. (eds.), Fl. Bras. 3(1): 39 (1842). Panama to S. Trop. America. 80 PAN 82 FRG GUY SUR VEN 83 CLM PER 84 BZC BZN. Cl. tuber geophyte.

var. *amazonum*
Panama to S. Trop. America. 80 PAN 82 FRG GUY SUR VEN 83 CLM PER 84 BZC BZN. Cl. tuber geophyte.
*Dioscorea cuspidata* Klotzsch ex Kunth, Enum. Pl. 5: 429 (1850), nom. inval.
*Dioscorea elegantula* Kunth, Enum. Pl. 5: 366 (1850).
*Dioscorea megalobotrya* Kunth & R.H.Schomb. in K.S.Kunth, Enum. Pl. 5: 365 (1850).
*Helmia consanguinea* Kunth, Enum. Pl. 5: 428 (1850). *Dioscorea amazonum* var. *consanguinea* (Kunth) Uline ex R.Knuth, Notizbl. Bot. Gart. Berlin-Dahlem 7: 217 (1917).
*Dioscorea amazonum* var. *burchellii* Uline ex R.Knuth, Notizbl. Bot. Gart. Berlin-Dahlem 7: 217 (1917).
*Dioscorea amazonum* var. *robustior* Uline ex R.Knuth, Notizbl. Bot. Gart. Berlin-Dahlem 7: 217 (1917).
*Dioscorea amazonum* var. *sagotiana* Uline ex R.Knuth, Notizbl. Bot. Gart. Berlin-Dahlem 7: 217 (1917).
*Dioscorea amazonum* var. *sprucei* Uline ex R.Knuth, Notizbl. Bot. Gart. Berlin-Dahlem 7: 217 (1917).
*Dioscorea surinamensis* Miq. ex Knuth in H.G.A.Engler (ed.), Pflanzenr., IV, 43: 147 (1924), nom. inval.
*Dioscorea georgensis* R.Knuth, Repert. Spec. Nov. Regni Veg. 28: 83 (1930).
*Dioscorea huberi* R.Knuth, Repert. Spec. Nov. Regni Veg. 28: 87 (1930).
*Dioscorea hitchcockii* R.Knuth, Repert. Spec. Nov. Regni Veg. 38: 118 (1935).
*Dioscorea holtii* R.Knuth, Repert. Spec. Nov. Regni Veg. 42: 163 (1937).

var. *klugii* (R.Knuth) Ayala, Diosc. Peru: 16 (1998). Peru. 83 PER. Cl. tuber geophyte.
*Dioscorea klugii* R.Knuth, Repert. Spec. Nov. Regni Veg. 30: 158 (1932).

**Dioscorea amoena** R.Knuth in H.G.A.Engler (ed.), Pflanzenr., IV, 43: 67 (1924). Venezuela. 82 VEN. Cl. tuber geophyte.

**Dioscorea analalavensis** Jum. & H.Perrier, Ann. Inst. Bot.-Géol. Colon. Marseille, II, 8: 399 (1910). N. Madagascar. 29 MDG. Tuber geophyte.

**Dioscorea ancachsensis** R.Knuth in H.G.A.Engler (ed.), Pflanzenr., IV, 43: 202 (1924). Peru. 83 PER. Tuber geophyte.

**Dioscorea andina** Phil., Anales Univ. Chile 93: 16 (1896). SC. Chile. 85 CLC. Cl. tuber geophyte.

**Dioscorea andromedusae** O.Téllez, Brittonia 48: 103 (1996). Peru. 83 PER. Cl. tuber geophyte.

**Dioscorea anomala** Griseb. in C.F.P.von Martius & auct. suc. (eds.), Fl. Bras. 3(1): 31 (1842). *Helmia anomala* (Griseb.) Kunth, Enum. Pl. 5: 427 (1850). SE. Brazil (Cadeia do Espinhaço). 84 BZL. Cl. tuber geophyte.

**Dioscorea antaly** Jum. & H.Perrier, Compt. Rend. Hebd. Séances Acad. Sci. 149: 485 (1909). Madagascar. 29 MDG. Cl. tuber geophyte.

**Dioscorea antucoana** Uline ex R.Knuth, Notizbl. Bot. Gart. Berlin-Dahlem 7: 207 (1917). SC. Chile. 85 CLC. Cl. tuber geophyte.

**Dioscorea arachidna** Prain & Burkill, J. Proc. Asiat. Soc. Bengal 10: 21 (1914). Assam to SC. China. 36 CHC 40 ASS 41 MYA THA VIE. Cl. tuber geophyte.
*Dioscorea collinsae* Prain & Burkill, Bull. Misc. Inform. Kew 1927: 234 (1927).

**Dioscorea araucana** Phil., Anales Univ. Chile 1873: 540 (1873). SC. Chile. 85 CLC. Cl. tuber geophyte.

**Dioscorea arcuatinervis** Hochr., Annuaire Conserv. Jard. Bot. Genève 11-12: 52 (1908). E. Madagascar. 29 MDG. Cl. tuber geophyte.
*Dioscorea humblotii* R.Knuth in H.G.A.Engler (ed.), Pflanzenr., IV, 43: 70 (1924).

**Dioscorea argyrogyna** Uline ex R.Knuth, Notizbl. Bot. Gart. Berlin-Dahlem 7: 204 (1917). WC. Brazil. 84 BZC. Cl. tuber geophyte.

**Dioscorea arifolia** C.Presl, Reliq. Haenk. 1: 134 (1827). Peru. 83 PER. Cl. tuber geophyte.

**Dioscorea aristolochiifolia** Poepp., Fragm. Syn. Pl.: 11 (1833). C. Chile. 85 CLC. Cl. tuber geophyte.
*Dioscorea trichoneura* Phil., Anales Univ. Chile 93: 6 (1896).

**Dioscorea asclepiadea** Prain & Burkill, Bull. Misc. Inform. Kew 1916: 190 (1916). Japan (Kyushu). 38 JAP. Cl. rhizome geophyte.

**Dioscorea aspera** Humb. & Bonpl. ex Willd., Sp. Pl. 4: 794 (1806). *Helmia aspera* (Humb. & Bonpl. ex Willd.) Kunth, Enum. Pl. 5: 429 (1850). *Dioscorea scabra* var. *aspera* (Humb. & Bonpl. ex Willd.) Uline ex R.Knuth, Notizbl. Bot. Gart. Berlin-Dahlem 7: 192 (1917). S. Venezuela, SE. & S. Brazil. 82 VEN 84 BZL BZS. Cl. tuber geophyte.

**Dioscorea aspersa** Prain & Burkill, J. Proc. Asiat. Soc. Bengal 4: 447 (1908). China (E. Yunnan, W. Guizhou). 36 CHC. Cl. tuber geophyte.
*Dioscorea pulverea* Prain & Burkill, J. Proc. Asiat. Soc. Bengal 10: 31 (1914).

**Dioscorea asperula** Pedralli, Napaea 8: 29 (1992). Brazil (Goiás, Minas Gerais). 84 BZC BZL. Cl. tuber geophyte.

**Dioscorea asteriscus** Burkill, Bull. Jard. Bot. État 15: 356 (1939). Kenya to Namibia. 23 CAF RWA ZAI 25 KEN TAN UGA? 26 MLW MOZ ZAM ZIM 27 BOT NAM. Cl. tuber geophyte.

**Dioscorea atrescens** R.Knuth in H.G.A.Engler (ed.), Pflanzenr., IV, 43: 245 (1924).

S. Venezuela. 82 VEN. Cl. tuber geophyte.

***Dioscorea auriculata*** Poepp., Fragm. Syn. Pl.: 13 (1833).
C. & S. Chile. 85 CLC CLS. Cl. tuber geophyte.
*Dioscorea helicifolia* Kunth, Enum. Pl. 5: 348 (1850).
*Dioscorea acutifolia* Phil., Linnaea 29: 64 (1857).

***Dioscorea bahiensis*** R.Knuth in H.G.A.Engler (ed.),
Pflanzenr., IV, 43: 351 (1924).
Brazil (Bahia). 84 BZE. Cl. tuber geophyte.

***Dioscorea balcanica*** Košanin, Oesterr. Bot. Z. 64: 37
(1914).
Montenegro to N. Albania. 13 ALB YUG. Cl. rhizome
geophyte.

***Dioscorea bancana*** Prain & Burkill, Bull. Misc. Inform.
Kew 1925: 62 (1925).
Sumatera (Bangka). 42 SUM. Cl. tuber geophyte.

***Dioscorea banzhuana*** S.J.Pei & C.T.Ting, Acta
Phytotax. Sin. 14(1): 70 (1976).
China (SE. Yunnan). 36 CHC. Cl. tuber geophyte.

***Dioscorea bartlettii*** C.V.Morton, Publ. Carnegie Inst.
Wash. 461: 242 (1936).
Mexico to C. America. 79 MXG MXS MXT 80 BLZ
GUA HON. Cl. tuber geophyte.

***Dioscorea basiclavicaulis*** Rizzini & A.Mattos, Revista
Brasil. Biol. 46: 317 (1986).
Brazil (E. Bahia). 84 BZE. Succ. rhizome geophyte.

***Dioscorea baya*** De Wild., Ann. Mus. Congo Belge, Bot.,
V, 3: 357 (1912).
Trop. Africa. 22 IVO LBR 23 BUR CAF CMN CON
GAB ZAI 25 UGA 26 ANG ZAM. Cl. tuber
geophyte.

var. ***baya***
C. Trop. Africa. 23 BUR CAF CMN CON GAB
ZAI 25 UGA 26 ANG ZAM. Cl. tuber geophyte.
*Dioscorea baya* var. *subcordata* De Wild., Bull.
Jard. Bot. État 4: 328 (1914).

var. ***kimpundi*** De Wild., Bull. Jard. Bot. État 4: 328
(1914).
W. & WC. Trop. Africa. 22 IVO LBR 23 CMN
CON ZAI. Cl. tuber geophyte.

***Dioscorea beecheyi*** R.Knuth in H.G.A.Engler (ed.),
Pflanzenr., IV, 43: 58 (1924).
S. Brazil. 84 BZS. Cl. tuber geophyte.

***Dioscorea belophylla*** (Prain) Voigt ex Haines, Forest Fl.
Chota Nagpur: 530 (1910).
Himalaya to Assam, S. India. 40 ASS BAN EHM IND
NEP WHM. Cl. tuber geophyte.
*Dioscorea sagittata* Royle, Ill. Bot. Himal. Mts.: 378
(1839), nom. inval.
*Dioscorea nummularia* var. *belophylla* Prain, Bengal
Pl. 2: 1065 (1903).

***Dioscorea bemandry*** Jum. & H.Perrier, Compt. Rend.
Hebd. Séances Acad. Sci. 149: 484 (1909).
NW. Madagascar. 29 MDG. Tuber geophyte.

***Dioscorea bemarivensis*** Jum. & H.Perrier, Ann. Inst.
Bot.-Géol. Colon. Marseille, II, 8: 423 (1910).
Aldabra, Comores, W. & S. Madagascar. 29 ALD
COM MDG. Cl. tuber geophyte.
*Dioscorea lucida* Scott-Elliot, J. Linn. Soc., Bot. 29:
60 (1891), nom. illeg.
*Dioscorea nesiotis* Hemsl., J. Bot. 55: 288 (1917).
*Dioscorea madagascariensis* R.Knuth in H.G.A.
Engler (ed.), Pflanzenr., IV, 43: 311 (1924).

*Dioscorea majungensis* R.Knuth in H.G.A.Engler
(ed.), Pflanzenr., IV, 43: 312 (1924).

***Dioscorea benthamii*** Prain & Burkill, J. Proc. Asiat.
Soc. Bengal 4: 448 (1908).
SE. China, Taiwan. 36 CHC 38 TAI. Cl. tuber
geophyte.
*Dioscorea tarokoensis* Hayata, Icon. Pl. Formosan. 10:
44 (1921).

***Dioscorea berenicea*** McVaugh, in Fl. Novo-Galiciana
15: 362 (1989).
Mexico (Jalisco). 79 MXS. Cl. tuber geophyte.

***Dioscorea bermejensis*** R.Knuth, Notizbl. Bot. Gart.
Berlin-Dahlem 7: 199 (1917).
S. Bolivia to NW. Argentina. 83 BOL 85 AGW. Cl.
tuber geophyte.

***Dioscorea bernoulliana*** Prain & Burkill, Bull. Misc.
Inform. Kew 1916: 192 (1916).
SE. Mexico to Guatemala. 79 MXT 80 GUA. Cl. tuber
geophyte.

***Dioscorea besseriana*** Kunth, Enum. Pl. 5: 345 (1850).
N. & C. Chile. 85 CLC CLN. Cl. tuber geophyte.
*Dioscorea besseriana* var. *berteroi* Uline ex R.Knuth,
Notizbl. Bot. Gart. Berlin-Dahlem 7: 205 (1917).

***Dioscorea beyrichii*** R.Knuth, Notizbl. Bot. Gart. Berlin-
Dahlem 7: 210 (1917).
Brazil (Rio de Janeiro). 84 BZL. Cl. tuber geophyte.

***Dioscorea bicolor*** Prain & Burkill, J. Proc. Asiat. Soc.
Bengal 4: 449 (1908).
China (SW. Sichuan, N. Yunnan). 36 CHC. Cl. tuber
geophyte.

***Dioscorea biformifolia*** S.J.Pei & C.T.Ting, Acta
Phytotax. Sin. 14(1): 69 (1976).
China (Yunnan). 36 CHC. Cl. rhizome geophyte.

***Dioscorea biloba*** (Phil.) Caddick & Wilkin, Taxon 51:
112 (2002).
N. Chile. 85 CLN. Tuber geophyte.
*Epipetrum bilobum* Phil., Anales Mus. Nac. Santiago
de Chile 2: 11 (1892).

***Dioscorea biplicata*** R.Knuth in H.G.A.Engler (ed.),
Pflanzenr., IV, 43: 77 (1924).
Colombia. 83 CLM. Cl. tuber geophyte.

***Dioscorea birmanica*** Prain & Burkill, J. Asiat. Soc.
Bengal, Pt. 2, Nat. Hist. 73: 185 (1904).
China (NW. Yunnan) to W. Indo-China. 36 CHC 41
MYA THA. Cl. rhizome geophyte.
*Dioscorea rangunensis* R.Knuth in H.G.A.Engler
(ed.), Pflanzenr., IV, 43: 320 (1924).

***Dioscorea birschelii*** Harms ex R.Knuth, Notizbl. Bot.
Gart. Berlin-Dahlem 7: 193 (1917).
N. Venezuela. 82 VEN. Cl. tuber geophyte.
*Dioscorea caracasensis* R.Knuth, Notizbl. Bot. Gart.
Berlin-Dahlem 7: 217 (1917), nom. illeg.

***Dioscorea blumei*** Prain & Burkill, J. Proc. Asiat. Soc.
Bengal 10: 25 (1914).
W. Jawa. 42 JAW SUM?. Cl. tuber geophyte.

***Dioscorea bolivarensis*** Steyerm., Fieldiana, Bot. 28(1):
158 (1951).
S. Venezuela. 82 VEN. Cl. tuber geophyte.

***Dioscorea bonii*** Prain & Burkill, Bull. Misc. Inform.
Kew 1933: 244 (1933).
N. Vietnam. 41 VIE. Cl. tuber geophyte.

**Dioscorea bosseri** Haigh & Wilkin, Kew Bull. 60: 276 (2005).
Madagascar. 29 MDG.

**Dioscorea brachybotrya** Poepp., Fragm. Syn. Pl.: 12 (1833).
C. & S. Chile, Argentina (Neuquén). 85 AGS CLC CLS. Cl. tuber geophyte.
*Dioscorea scandens* Kunze ex Kunth, Enum. Pl. 5: 350 (1850).
*Dioscorea novemloba* Steud. ex F.Phil., Cat. Pl. Vasc. Chil.: 286 (1881), nom. nud.
*Dioscorea brachybotrya* var. *germainii* Uline ex R.Knuth, Notizbl. Bot. Gart. Berlin-Dahlem 7: 206 (1917).
*Dioscorea buchtienii* R.Knuth, Notizbl. Bot. Gart. Berlin-Dahlem 7: 207 (1917).

**Dioscorea brachystachya** Phil., Anales Univ. Chile 93: 7 (1896).
SC. Chile. 85 CLC. Cl. tuber geophyte.

**Dioscorea bradei** R.Knuth, Repert. Spec. Nov. Regni Veg. 42: 177 (1937).
Brazil (Rio de Janeiro). 84 BZL. Cl. tuber geophyte.

**Dioscorea brandisii** Prain & Burkill, J. Proc. Asiat. Soc. Bengal 10: 27 (1914).
Myanmar. 41 MYA. Cl. tuber geophyte.

**Dioscorea brevipetiolata** Prain & Burkill, J. Proc. Asiat. Soc. Bengal 10: 38 (1914).
Indo-China. 41 CBD THA VIE. Cl. tuber geophyte.

**Dioscorea bridgesii** Griseb. ex Kunth, Enum. Pl. 5: 358 (1850).
C. Chile. 85 CLC. Cl. tuber geophyte.
*Rajania flexuosa* Poir. in J.B.A.M.de Lamarck, Encycl. 6: 59 (1804).

**Dioscorea brownii** Schinz, Mém. Herb. Boissier 20: 11 (1900).
Cape Prov. to KwaZulu-Natal. 27 CPP NAT. Cl. tuber geophyte.

**Dioscorea bryoniifolia** Poepp., Fragm. Syn. Pl.: 13 (1833).
C. Chile. 85 CLC. Cl. tuber geophyte.
*Dioscorea hederacea* Miers, Trav. Chile 2: 531 (1826), nom. nud.

**Dioscorea buchananii** Benth., Hooker's Icon. Pl. 14: t. 1397 (1879).
Tanzania to S. Trop. Africa. 23 ZAI 25 TAN 26 ANG MLW MOZ ZAM ZIM. Cl. tuber geophyte.
*Dioscorea buchananii* var. *ukamensis* R.Knuth in H.G.A.Engler (ed.), Pflanzenr., IV, 43: 185 (1924).
*Dioscorea mildbraediana* R.Knuth, Notizbl. Bot. Gart. Berlin-Dahlem 11: 1059 (1934).
*Dioscorea rhacodes* Peter ex R.Knuth, Repert. Spec. Nov. Regni Veg. 42: 162 (1937).

**Dioscorea bulbifera** L., Sp. Pl.: 1033 (1753). *Helmia bulbifera* (L.) Kunth, Enum. Pl. 5: 435 (1850). *Polynome bulbifera* (L.) Salisb., Gen. Pl.: 12 (1866).
Trop. & Subtrop. Old World. 22 BEN BKN GHA GUI IVO LBR NGA SEN SIE TOG 23 CAF CMN GAB RWA ZAI 24 CHA ERI ETH SUD 25 KEN TAN UGA 26 MLW MOZ ZAM ZIM 27 CPV 29 MDG sey 36 CHC CHH CHN CHS CHT 38 JAP KOR kzn NNS oga TAI 40 ASS BAN EHM IND NEP PAK SRL 41 AND CBD MYA THA VIE 42 BOR JAW LSI MLY PHI SUL SUM 43 NWG 50 NTA QLD WAU (60) fij nue ton (61) mrq sci 62 CRL (63) haw (78) fla (79) (80) cos (81) jam lee pue trt win (82) frg guy sur ven (83) per (84) bzc bze bzl. Cl. tuber geophyte.
*Dioscorea sylvestris* De Wild. in ?.
*Dioscorea tamifolia* Salisb., Parad. Lond.: t. 17 (1805).
*Dioscorea crispata* Roxb., Fl. Ind. ed. 1832, 3: 802 (1832). *Dioscorea bulbifera* var. *crispata* (Roxb.) Prain, Bengal Pl. 2: 1066 (1903).
*Dioscorea heterophylla* Roxb., Fl. Ind. ed. 1832, 3: 804 (1832).
*Dioscorea pulchella* Roxb., Fl. Ind. ed. 1832, 3: 801 (1832). *Dioscorea bulbifera* var. *pulchella* (Roxb.) Prain, Bengal Pl. 2: 1066 (1903).
*Dioscorea tenuiflora* Schltdl., Linnaea 17: 608 (1843).
*Dioscorea latifolia* Benth. in W.J.Hooker, Niger Fl.: 535 (1849).
*Dioscorea hoffa* Cordem., Fl. Réunion: 159 (1895).
*Dioscorea sativa* var. *elongata* F.M.Bailey, Queensl. Fl. 5: 1615 (1902). *Dioscorea bulbifera* var. *elongata* (F.M.Bailey) Prain & Burkill, J. Proc. Asiat. Soc. Bengal 10: 27 (1914).
*Dioscorea sativa* var. *rotunda* F.M.Bailey, Queensl. Fl. 5: 1615 (1902).
*Dioscorea bulbifera* var. *sativa* Prain, Bengal Pl. 2: 1066 (1903).
*Dioscorea hofika* Jum. & H.Perrier, Ann. Inst. Bot.-Géol. Colon. Marseille, II, 8: 291 (1910).
*Dioscorea sativa* f. *domestica* Makino in Y.Iinuma, Somoku-Dzusetsu, ed. 3, 4: 1326 (1912).
*Dioscorea anthropophagorum* A.Chev., Veg. Ut. Afr. Trop. Franç. 8: 357 (1913). *Dioscorea bulbifera* var. *anthropophagorum* (A.Chev.) Summerh. in ?.
*Dioscorea longipetiolata* Baudon, Ann. Inst. Bot.-Géol. Colon. Marseille, III, 1: 242 (1913).
*Dioscorea violacea* Baudon, Ann. Inst. Bot.-Géol. Colon. Marseille, III, 1: 242 (1913), nom. illeg.
*Dioscorea bulbifera* var. *suavia* Prain & Burkill, J. Proc. Asiat. Soc. Bengal 10: 26 (1914).
*Dioscorea bulbifera* var. *vera* Prain & Burkill, J. Proc. Asiat. Soc. Bengal 10: 26 (1914).
*Dioscorea rogersii* Prain & Burkill, J. Proc. Asiat. Soc. Bengal 10: 27 (1914).
*Dioscorea korrorensis* R.Knuth in H.G.A.Engler (ed.), Pflanzenr., IV, 43: 190 (1924).
*Dioscorea perrieri* R.Knuth in H.G.A.Engler (ed.), Pflanzenr., IV, 43: 354 (1924).

**Dioscorea bulbotricha** Hand.-Mazz., Denkschr. Kaiserl. Akad. Wiss., Wien. Math.-Naturwiss. Kl. 79: 221 (1908).
SE. Brazil. 84 BZL. Cl. tuber geophyte.

**Dioscorea burchellii** Baker, J. Bot. 27: 1 (1889).
Cape Prov. 27 CPP. Cl. tuber geophyte.

**Dioscorea burkilliana** J.Miège, Bull. Inst. Fondam. Afrique Noire, Sér. A., Sci. Nat. 20: 48 (1958).
W. & WC. Trop. Africa. 22 BEN IVO SIE 23 GAB. Cl. tuber geophyte.

**Dioscorea cachipuertensis** Ayala, Diosc. Peru: 19 (1998).
Peru. 83 PER. Cl. tuber geophyte.

**Dioscorea calcicola** Prain & Burkill, Bull. Misc. Inform. Kew 1925: 64 (1925).
Pen. Thailand to N. Pen. Malaysia. 41 THA 42 MLY. Cl. tuber geophyte.

**Dioscorea caldasensis** R.Knuth in H.G.A.Engler (ed.), Pflanzenr., IV, 43: 112 (1924).
Brazil (Minas Gerais). 84 BZL. Cl. tuber geophyte.

*Dioscorea fimbriata* Uline ex R.Knuth, Notizbl. Bot. Gart. Berlin-Dahlem 7: 196 (1917), nom. illeg.

**Dioscorea calderillensis** R.Knuth, Notizbl. Bot. Gart. Berlin-Dahlem 7: 197 (1917). S. Bolivia. 83 BOL. Cl. tuber geophyte.

**Dioscorea callacatensis** R.Knuth, Repert. Spec. Nov. Regni Veg. 28: 83 (1930). Peru. 83 PER. Cl. tuber geophyte.

**Dioscorea cambodiana** Prain & Burkill, J. Proc. Asiat. Soc. Bengal 10: 12 (1914). Indo-China. 41 CBD VIE. Cl. tuber geophyte.

**Dioscorea campanulata** Uline ex R.Knuth, Notizbl. Bot. Gart. Berlin-Dahlem 7: 189 (1917). Brazil (Rio de Janeiro). 84 BZL. Cl. tuber geophyte.
  *Dioscorea campanulata* var. *lanceolata* Uline, Notizbl. Bot. Gart. Berlin-Dahlem 7: 189 (1917).

**Dioscorea campestris** Griseb. in C.F.P.von Martius & auct. suc. (eds.), Fl. Bras. 3(1): 30 (1842). *Helmia campestris* (Griseb.) Kunth, Enum. Pl. 5: 425 (1850). Brazil to N. Argentina. 84 BZC BZE BZL BZN BZS 85 AGE AGW PAR. Cl. tuber geophyte.
  *Dioscorea campestris* var. *grandiflora* Griseb. in C.F.P.von Martius & auct. suc. (eds.), Fl. Bras. 3(1): 30 (1842).
  *Dioscorea campestris* var. *parviflora* Griseb. in C.F.P.von Martius & auct. suc. (eds.), Fl. Bras. 3(1): 30 (1842).
  *Dioscorea campestris* f. *longispicata* Hauman, Anales Mus. Nac. Hist. Nat. Buenos Aires 27: 451 (1915).
  *Dioscorea campestris* var. *longispicata* Hauman, Anales Mus. Nac. Hist. Nat. Buenos Aires 27: 451 (1916).
  *Dioscorea campestris* f. *pedalis* Uline ex R.Knuth, Notizbl. Bot. Gart. Berlin-Dahlem 7: 187 (1917).
  *Dioscorea campestris* f. *piedadensis* Uline ex R.Knuth, Notizbl. Bot. Gart. Berlin-Dahlem 7: 186 (1917).
  *Dioscorea campestris* f. *plantaginifolia* Uline ex R.Knuth, Notizbl. Bot. Gart. Berlin-Dahlem 7: 186 (1917).
  *Dioscorea campestris* f. *stenorachis* Uline ex R.Knuth, Notizbl. Bot. Gart. Berlin-Dahlem 7: 186 (1917).
  *Dioscorea campestris* f. *paraguayensis* R.Knuth, Repert. Spec. Nov. Regni Veg. 22: 346 (1926).

**Dioscorea campos-portoi** R.Knuth, Repert. Spec. Nov. Regni Veg. 42: 176 (1937). Brazil (Rio de Janeiro). 84 BZL. Cl. tuber geophyte.

**Dioscorea carionis** Prain & Burkill, Bull. Misc. Inform. Kew 1916: 193 (1916). SE. Mexico to Guatemala. 79 MXT 80 GUA. Cl. tuber geophyte.

**Dioscorea carpomaculata** O.Téllez & B.G.Schub., Ann. Missouri Bot. Gard. 78: 245 (1991). Mexico to C. America. 79 MXC MXE MXN MXS MXT 80 ELS GUA. Cl. tuber geophyte.

var. **carpomaculata**. Mexico to C. America. 79 MXC MXE MXN MXS MXT 80 ELS GUA. Cl. tuber geophyte.

var. **cinerea** (Uline ex R.Knuth) O.Téllez & B.G. Schub., Ann. Missouri Bot. Gard. 78: 248 (1991). S. Mexico. 79 MXS MXT. Cl. tuber geophyte.
  *Dioscorea cymosula* var. *cinerea* Uline ex R.Knuth, Notizbl. Bot. Gart. Berlin-Dahlem 7: 202 (1917).

**Dioscorea castilloniana** Hauman, Bol. Mus. Hist. Nat., Tucuman 11: 31 (1927). Argentina (Jujuy). 85 AGW. Cl. tuber geophyte.

**Dioscorea catharinensis** R.Knuth, Notizbl. Bot. Gart. Berlin-Dahlem 7: 201 (1917). SE. & S. Brazil. 84 BZL BZS. Cl. tuber geophyte.
  *Dioscorea catharinensis* R.Knuth, Notizbl. Bot. Gart. Berlin-Dahlem 7: 213 (1917), nom. illeg.

**Dioscorea caucasica** Lipsky, Zap. Kievsk. Obshch. Estestvoissp. 1893: 143 (1893). Transcaucasus. 33 TCS. Cl. rhizome geophyte.

**Dioscorea cayennensis** Lam., Encycl. 3: 233 (1789). W. Trop. Africa to Cameroon, introduced elsewere. 22 BEN BKN GAM GHA GNB GUI IVO LBR MLI NGA NGR SEN SIE TOG 23 CMN (80) cos (81) jam lee pue trt win (82) frg guy sur ven (84) bzl bzn. Cl. tuber geophyte.

subsp. **cayennensis** W. Trop. Africa to Cameroon, introduced elsewere. 22 BEN BKN GHA GNB GUI IVO LBR NGA NGR SEN SIE TOG 23 CMN (80) cos (81) jam lee pue trt win (82) frg guy sur ven (84) bzl bzn. Cl. tuber geophyte.
  *Dioscorea aculeata* Balb. ex Kunth, Enum. Pl. 5: 381 (1850), nom. illeg.
  *Dioscorea berteroana* Kunth, Enum. Pl. 5: 381 (1850).
  *Dioscorea moma* De Wild., Ann. Mus. Congo Belge, Bot., V, 3: 367 (1912).
  *Dioscorea pruinosa* A.Chev., Études Fl. Afr. Centr. Franç.: 311 (1913), nom. nud.
  *Dioscorea camerunensis* R.Knuth in H.G.A.Engler (ed.), Pflanzenr., IV, 43: 298 (1924).
  *Dioscorea occidentalis* R.Knuth in H.G.A.Engler (ed.), Pflanzenr., IV, 43: 299 (1924).

subsp. **rotundata** (Poir.) J.Miège in J.Hutchinson & J.M.Dalziel, Fl. W. Trop. Afr., ed. 2, 3(1): 153 (1968). W. Trop. Africa. 22 BEN BKN GAM GHA GNB GUI IVO LBR MLI NGA NGR SEN SIE TOG (81) jam. Cl. tuber geophyte.
  *Dioscorea rotundata* Poir. in J.B.A.M.de Lamarck, Encycl., Suppl. 3: 139 (1813).

**Dioscorea ceratandra** Uline ex R.Knuth, Notizbl. Bot. Gart. Berlin-Dahlem 7: 207 (1917). Peru to N. Argentina. 83 BOL PER 84 BZL 85 AGE AGW PAR. Cl. tuber geophyte.
  *Dioscorea congestiflora* R.Knuth, Notizbl. Bot. Gart. Berlin-Dahlem 7: 208 (1917).
  *Dioscorea quirogae* R.Knuth in H.G.A.Engler (ed.), Pflanzenr., IV, 43: 218 (1924).

**Dioscorea chacoensis** R.Knuth, Repert. Spec. Nov. Regni Veg. 21: 77 (1925). Bolivia. 83 BOL. Cl. tuber geophyte.

**Dioscorea chagllaensis** R.Knuth, Repert. Spec. Nov. Regni Veg. 28: 88 (1930). Peru. 83 PER. Cl. tuber geophyte.

**Dioscorea chancayensis** R.Knuth, Repert. Spec. Nov. Regni Veg. 28: 81 (1930). Peru. 83 PER. Cl. tuber geophyte.

**Dioscorea chaponensis** R.Knuth, Repert. Spec. Nov. Regni Veg. 38: 117 (1935). Costa Rica to Peru. 80 COS PAN 83 CLM PER. Cl. tuber geophyte.

var. **chaponensis** Costa Rica to N. Colombia. 80 COS PAN 83 CLM. Cl. tuber geophyte.

var. *chazutensis* Ayala, Diosc. Peru: 22 (1998).
Peru. 83 PER. Cl. tuber geophyte.

**Dioscorea chiapasensis** Matuda, Bol. Soc. Bot. México
15: 25 (1953).
SE. Mexico to Guatemala. 79 MXT 80 GUA. Cl. tuber
geophyte.

**Dioscorea chimborazensis** R.Knuth, Repert. Spec. Nov.
Regni Veg. 28: 86 (1930).
Ecuador. 83 ECU. Cl. tuber geophyte.

**Dioscorea chingii** Prain & Burkill, Bull. Misc. Inform.
Kew 1931: 425 (1931).
China (Yunnan, Guangxi) to Vietnam. 36 CHC CHS
41 VIE. Cl. rhizome geophyte.

**Dioscorea choriandra** Uline ex R.Knuth, Notizbl. Bot.
Gart. Berlin-Dahlem 7: 198 (1917).
Ecuador. 83 ECU. Cl. tuber geophyte.

**Dioscorea chouardii** Gaussen, Bull. Soc. Bot. France 99:
24 (1952). *Borderea chouardii* (Gaussen) Gaussen &
Heslot, Bull. Soc. Hist. Nat. Toulouse: 390 (1965).
C. Pyrenees (Valley of the Rio Noguera Ribagorzana).
12 SPA. Tuber geophyte.

**Dioscorea cienegensis** R.Knuth, Notizbl. Bot. Gart.
Berlin-Dahlem 7: 200 (1917).
N. Argentina. 85 AGE AGW. Cl. tuber geophyte.

**Dioscorea cinnamomifolia** Hook., Bot. Mag. 55: t. 2825
(1828).
E. & S. Brazil. 84 BZE BZL BZS. Cl. tuber geophyte.
*Dioscorea tuberosa* Vell., Fl. Flumin. 10: t. 125 (1831).
*Dioscorea teretiuscula* Griseb. in C.F.P.von Martius &
auct. suc. (eds.), Fl. Bras. 3(1): 48 (1842).
*Dioscorea zanoniae* Klotzsch ex Griseb. in C.F.P.von
Martius & auct. suc. (eds.), Fl. Bras. 3(1): 38
(1842). *Dioscorea cinnamomifolia* var. *zanoniae*
(Klotzsch ex Griseb.) R.Knuth in H.G.A.Engler
(ed.), Pflanzenr., IV, 43: 64 (1924).
*Rajania brasiliensis* Griseb. in C.F.P.von Martius &
auct. suc. (eds.), Fl. Bras. 3(1): 48 (1842).

**Dioscorea cirrhosa** Lour., Fl. Cochinch.: 625 (1790).
*Strophis cirrhosa* (Lour.) Salisb., Gen. Pl.: 12 (1866).
E. China to N. Thailand and Nansei-shoto. 36 CHH
CHH CHS CHT 38 NNS TAI 41 THA VIE. Cl.
tuber geophyte.

var. *cirrhosa*
E. China to N. Thailand and Nansei-shoto. 36 CHH
CHS CHT 38 NNS TAI 41 THA VIE. Cl. tuber
geophyte.
*Dioscorea rhipogonoides* Oliv., Hooker's Icon. Pl.
19: t. 1868 (1889).
*Dioscorea bonnetii* A.Chev., Bull. Écon. Indochine,
n.s., 20: 328 (1918).
*Dioscorea matsudae* Hayata, Icon. Pl. Formosan.
10: 39 (1921).
*Dioscorea angusta* R.Knuth in H.G.A.Engler (ed.),
Pflanzenr., IV, 43: 288 (1924).
*Dioscorea camphorifolia* Uline ex R.Knuth in H.G.A.
Engler (ed.), Pflanzenr., IV, 43: 288 (1924).

var. *cylindrica* C.T.Ting & M.C.Chang, Acta
Phytotax. Sin. 20: 206 (1982).
Hainan. 36 CHH. Cl. tuber geophyte.

**Dioscorea cissophylla** Phil., Anales Univ. Chile 93: 18
(1896).
Chile (I. de Chiloé). 85 CLS. Cl. tuber geophyte.

**Dioscorea claessensii** De Wild., Ann. Mus. Congo
Belge, Bot., V, 3: 358 (1912).
WC. Trop. Africa. 23 CMN ZAI. Cl. tuber geophyte.

**Dioscorea claussenii** Uline ex R.Knuth, Notizbl. Bot.
Gart. Berlin-Dahlem 7: 215 (1917).
Brazil (Minas Gerais, Goiás). 84 BZC BZL. Cl. tuber
geophyte.

**Dioscorea claytonii** Ayala, Ann. Missouri Bot. Gard. 68:
130 (1981).
Peru. 83 PER. Cl. tuber geophyte.

**Dioscorea cochleariapiculata** De Wild., Bull. Jard. Bot.
État 4: 350 (1914).
Ethiopia to S. Trop. Africa. 23 BUR ZAI 24 ETH 25
TAN 26 MLW MOZ ZAM ZIM 27 CPV NAM. Cl.
tuber geophyte.
*Dioscorea stolzii* R.Knuth in H.G.A.Engler (ed.),
Pflanzenr., IV, 43: 136 (1924).

**Dioscorea collettii** Hook.f., Fl. Brit. India 6: 290 (1892).
*Dioscorea gracillima* var. *collettii* (Hook.f.) Uline ex
R.Knuth in H.G.A.Engler (ed.), Pflanzenr., IV, 43:
253 (1924).
Indo-China to Japan. 36 CHC CHS 38 JAP TAI 41
LAO MYA THA VIE. Cl. rhizome geophyte.

var. *collettii*
Indo-China to Taiwan. 36 CHC CHS 38 TAI 41
LAO MYA THA VIE. Cl. rhizome geophyte.
*Dioscorea oenea* Prain & Burkill, J. Proc. Asiat.
Soc. Bengal 10: 16 (1914).
*Dioscorea kelungensis* Hayata, Icon. Pl. Formosan.
10: 36 (1921).
*Dioscorea tashiroi* Hayata, Icon. Pl. Formosan. 10:
44 (1921).
*Dioscorea nigrescens* R.Knuth in H.G.A.Engler
(ed.), Pflanzenr., IV, 43: 253 (1924), nom. illeg.
*Dioscorea hui* R.Knuth, Repert. Spec. Nov. Regni
Veg. 21: 80 (1925).
*Dioscorea seniavinii* Prain & Burkill, Bull. Misc.
Inform. Kew 1925: 59 (1925).

var. *hypoglauca* (Palib.) S.J.Pei & C.T.Ting, Acta
Phytotax. Sin. 14(1): 66 (1976).
S. China, Japan, N. Taiwan. 36 CHC CHS 38 JAP
TAI. Cl. rhizome geophyte.
*Dioscorea hypoglauca* Palib., Bull. Herb. Boissier,
II, 6: 2 (1906).
*Dioscorea morsei* Prain & Burkill, J. Proc. Asiat.
Soc. Bengal 4: 454 (1908).
*Dioscorea undulata* R.Knuth in H.G.A.Engler (ed.),
Pflanzenr., IV, 43: 315 (1924).
*Dioscorea kaoi* T.S.Liu & T.C.Huang, Taiwania 7:
33 (1960).
*Dioscorea izuensis* Akahori, Acta Phytotax. Geobot.
19: 161 (1963).

**Dioscorea communis** (L.) Caddick & Wilkin, Taxon 51:
112 (2002).
Macaronesia, W. Europe to Medit. and Iran. 10 GRB
ire 11 AUT BGM GER HUN SWI 12 BAL COR
FRA POR SAR SPA 13 ALB BUL GRC ITA KRI
ROM SIC TUE YUG 14 KRY 20 ALG LBY MOR
TUN 21 AZO CNY MDR 33 NCS TCS 34 CYP
EAI IRN IRQ LBS PAL TUR (51) nzn. Cl. tuber
geophyte.
*Tamus communis* L., Sp. Pl.: 1028 (1753).
*Tamus cretica* L., Sp. Pl.: 1028 (1753). *Tamus communis*
subsp. *cretica* (L.) Nyman, Consp. Fl. Eur.: 718

(1882). *Tamus communis* var. *cretica* (L.) Boiss., Fl. Orient. 5: 344 (1882).

*Tamus racemosa* Gouan, Fl. Monsp.: 426 (1765).

*Smilax rubra* Willd., Enum. Pl.: 1015 (1809).

*Tamus cordifolia* Stokes, Bot. Mat. Med. 4: 544 (1812).

*Tamus edulis* Lowe, Trans. Cambridge Philos. Soc. 4(1): 12 (1833).

*Tamus norsa* Lowe, Trans. Cambridge Philos. Soc. 6(3): 21 (1838).

*Tamus communis* var. *subtriloba* Guss., Fl. Sicul. Syn. 2: 880 (1844). *Tamus communis* f. *subtriloba* (Guss.) O.Bolòs & Vigo, Fl. Man. Paisos Catalans 4: 171 (2001).

*Dioscorea canariensis* Webb & Berthel., Hist. Nat. Iles Canaries 2(3): 317 (1847).

*Tamus canariensis* Willd. ex Kunth, Enum. Pl. 5: 455 (1850).

*Tamus parviflora* Kunth, Enum. Pl. 5: 454 (1850).

*Tamus communis* var. *triloba* Simonk., Enum. Fl. Transsilv.: 520 (1887).

*Tamus baccifera* St.-Lag. in A.Cariot, Étude Fl., ed. 8, 2: 765 (1889).

*Tamus cirrhosa* Hausskn. ex Bornm., Beih. Bot. Centralbl. 58 E: 271 (1938).

**Dioscorea commutata** R.Knuth in H.G.A.Engler (ed.), Pflanzenr., IV, 43: 74 (1924).
S. Brazil. 84 BZS. Cl. tuber geophyte.

**Dioscorea comorensis** R.Knuth, Repert. Spec. Nov. Regni Veg. 22: 347 (1926).
Comoros. 29 COM. Cl. tuber geophyte.

**Dioscorea composita** Hemsl., Biol. Cent.-Amer., Bot. 3: 354 (1884).
Mexico to C. America. 79 MXC MXG MXS MXT 80 BLZ COS ELS GUA HON. Cl. tuber geophyte.

*Dioscorea tepinapensis* Uline ex R.Knuth, Notizbl. Bot. Gart. Berlin-Dahlem 7: 204 (1917).

*Dioscorea tepinapensis* var. *aggregata* Uline ex R.Knuth, Notizbl. Bot. Gart. Berlin-Dahlem 7: 205 (1917).

**Dioscorea contracta** R.Knuth, Notizbl. Bot. Gart. Berlin-Dahlem 7: 190 (1917).
Brazil (Minas Gerais). 84 BZL. Cl. tuber geophyte.

**Dioscorea convolvulacea** Cham. & Schltdl., Linnaea 6: 49 (1831). *Helmia convolvulacea* (Cham. & Schltdl.) Kunth, Enum. Pl. 5: 415 (1850).
Mexico to C. America, Trinidad. 79 MXC MXE MXG MXS 80 BLZ COS ELS GUA NIC PAN 81 TRT. Cl. tuber geophyte.

subsp. **convolvulacea**
Mexico to C. America, Trinidad. 79 MXC MXE MXG MXS 80 BLZ COS ELS GUA NIC PAN 81 TRT. Cl. tuber geophyte.

*Dioscorea hirsuta* M.Martens & Galeotti, Bull. Acad. Roy. Sci. Bruxelles 9(2): 391 (1842), nom. illeg.

*Dioscorea macrostachya* M.Martens & Galeotti, Bull. Acad. Roy. Sci. Bruxelles 9(2): 892 (1842), nom. illeg.

*Dioscorea brachycarpa* Schltdl., Linnaea 17: 109 (1843). *Helmia brachycarpa* (Schltdl.) Kunth, Enum. Pl. 5: 416 (1850).

*Dioscorea capillaris* Hemsl., Biol. Cent.-Amer., Bot. 3: 354 (1884).

*Dioscorea capillaris* var. *glabra* Hemsl., Biol. Cent.-Amer., Bot. 3: 323 (1884). *Dioscorea convolvulacea* var. *glabra* (Hemsl.) Uline ex

R.Knuth in H.G.A.Engler (ed.), Pflanzenr., IV, 43: 99 (1924).

*Dioscorea esurientium* Uline, Bot. Jahrb. Syst. 22: 429 (1896). *Dioscorea convolvulacea* subsp. *esurientium* (Uline) Uline ex R.Knuth in H.G.A. Engler (ed.), Pflanzenr., IV, 43: 99 (1924).

*Dioscorea hirsuta* var. *glabra* Uline, Bot. Jahrb. Syst. 22: 428 (1896).

*Dioscorea joseensis* R.Knuth in H.G.A.Engler (ed.), Pflanzenr., IV, 43: 159 (1924).

*Dioscorea chamela* McVaugh, in Fl. Novo-Galiciana 15: 364 (1989).

subsp. **grandifolia** (Schltdl.) Uline ex R.Knuth in H.G.A.Engler (ed.), Pflanzenr., IV, 43: 99 (1924). Mexico. 79 MXS.

*\*Dioscorea grandifolia* Schltdl., Linnaea 17: 602 (1843).

**Dioscorea conzattii** R.Knuth, Repert. Spec. Nov. Regni Veg. 11: 221 (1936).
Mexico (Oaxaca). 79 MXS. Cl. tuber geophyte.

**Dioscorea cordifolia** Laness., Pl. Util. Col. Franç.: 413 (1886).
French Guiana. 82 FRG. Cl. tuber geophyte.

**Dioscorea coreana** (Prain & Burkill) R.Knuth in H.G.A. Engler (ed.), Pflanzenr., IV, 43: 175 (1924).
Korea. 38 KOR. Cl. rhizome geophyte.

*\*Dioscorea villosa* var. *coreana* Prain & Burkill, J. Proc. Asiat. Soc. Bengal 10: 15 (1914).

**Dioscorea coriacea** Humb. & Bonpl. ex Willd., Sp. Pl. 4: 794 (1806). *Helmia coriacea* (Humb. & Bonpl. ex Willd.) Kunth, Enum. Pl. 5: 422 (1850).
W. South America to Venezuela. 82 VEN 83 CLM ECU PER. Cl. tuber geophyte.

*Dioscorea frutescens* Rusby, Descr. S. Amer. Pl.: 5 (1920).

*Dioscorea pennellii* R.Knuth, Repert. Spec. Nov. Regni Veg. 22: 346 (1926).

*Dioscorea pennellii* var. *pilosula* R.Knuth, Repert. Spec. Nov. Regni Veg. 22: 346 (1926).

*Dioscorea caucensis* R.Knuth, Repert. Spec. Nov. Regni Veg. 28: 86 (1930).

**Dioscorea coripatensis** J.F.Macbr., Candollea 6: 2 (1934).
Bolivia. 83 BOL. Cl. tuber geophyte.

\* *Dioscorea glauca* Rusby, Bull. New York Bot. Gard. 4: 459 (1907), nom. illeg.

**Dioscorea coronata** Hauman, Anales Mus. Nac. Hist. Nat. Buenos Aires 27: 480 (1916).
S. Brazil to NE. Argentina. 84 BZS 85 AGE. Cl. tuber geophyte.

*Dioscorea acutata* R.Knuth in H.G.A.Engler (ed.), Pflanzenr., IV, 43: 125 (1924).

*Dioscorea praetervisa* R.Knuth, Bull. Misc. Inform. Kew 1925: 122 (1925).

**Dioscorea cotinifolia** Kunth, Enum. Pl. 5: 386 (1850).
S. Mozambique to S. Africa. 26 MOZ 27 BOT CPP NAT SWZ TVL. Cl. tuber geophyte.

*Dioscorea malifolia* Baker, J. Bot. 27: 1 (1889).

**Dioscorea craibiana** Prain & Burkill, Bull. Misc. Inform. Kew 1931: 425 (1931).
Thailand. 41 THA. Cl. tuber geophyte.

**Dioscorea crateriflora** R.Knuth, Repert. Spec. Nov. Regni Veg. 22: 344 (1926).
Venezuela. 82 VEN. Cl. tuber geophyte.

*Dioscorea crotalariifolia* Uline in H.G.A.Engler & K.A.E.Prantl, Nat. Pflanzenfam., Nachtr. 1: 85 (1897).
Guyana to Peru. 82 GUY VEN 83 PER 84 BZN. Cl. tuber geophyte.
*Dioscorea quinquefoliolata* R.Knuth, Repert. Spec. Nov. Regni Veg. 40: 222 (1936).

*Dioscorea cruzensis* R.Knuth, Notizbl. Bot. Gart. Berlin-Dahlem 7: 194 (1917).
Mexico (Veracruz). 79 MXG. Cl. tuber geophyte.

*Dioscorea cubensis* R.Knuth, Notizbl. Bot. Gart. Berlin-Dahlem 7: 209 (1917).
Cuba. 81 CUB. Cl. tuber geophyte.

*Dioscorea cumingii* Prain & Burkill, J. Proc. Asiat. Soc. Bengal 4: 449 (1908).
Taiwan (Lan Yü), Philippines. 38 TAI 42 PHI. Cl. tuber geophyte.
*Dioscorea elmeri* Prain & Burkill, Leafl. Philip. Bot. 5: 1594 (1913).
*Dioscorea inaequifolia* Elmer ex Prain & Burkill, Leafl. Philip. Bot. 5: 1595 (1913).
*Dioscorea echinata* R.Knuth in H.G.A.Engler (ed.), Pflanzenr., IV, 43: 148 (1924).
*Dioscorea polyphylla* R.Knuth in H.G.A.Engler (ed.), Pflanzenr., IV, 43: 148 (1924).
*Dioscorea heptaphylla* Sasaki, Trans. Nat. Hist. Soc. Taiwan 21: 147 (1931).

*Dioscorea curitybensis* R.Knuth in H.G.A.Engler (ed.), Pflanzenr., IV, 43: 113 (1924).
S. Brazil. 84 BZS. Cl. tuber geophyte.

*Dioscorea cuspidata* Humb. & Bonpl. ex Willd., Sp. Pl. 4: 794 (1806). *Helmia cuspidata* (Humb. & Bonpl. ex Willd.) Kunth, Enum. Pl. 5: 428 (1850).
SE. Colombia to S. Venezuela. 82 VEN 83 CLM. Cl. tuber geophyte.
*Dioscorea pichinchensis* R.Knuth in H.G.A.Engler (ed.), Pflanzenr., IV, 43: 57 (1924).

*Dioscorea cuyabensis* R.Knuth, Notizbl. Bot. Gart. Berlin-Dahlem 7: 216 (1917).
Brazil (Minas Gerais). 84 BZC. Cl. tuber geophyte.

*Dioscorea cyanisticta* J.D.Sm., Bot. Gaz. 20: 10 (1895).
SE. Mexico to C. America. 79 MXT 80 COS GUA NIC. Cl. tuber geophyte.

*Dioscorea cymosula* Hemsl., Biol. Cent.-Amer., Bot. 3: 355 (1884).
Costa Rica to Panama. 80 COS PAN. Cl. tuber geophyte.
*Dioscorea cymosula* var. *duchassaingii* Uline ex R.Knuth, Notizbl. Bot. Gart. Berlin-Dahlem 7: 203 (1917).
*Dioscorea cymosula* var. *longiracemosa* Uline ex R.Knuth, Notizbl. Bot. Gart. Berlin-Dahlem 7: 203 (1917).
*Dioscorea permollis* R.Knuth in H.G.A.Engler (ed.), Pflanzenr., IV, 43: 61 (1924).

*Dioscorea cyphocarpa* C.B.Rob. ex Knuth, Notizbl. Bot. Gart. Berlin-Dahlem 7: 209 (1917).
Mexico (Guerrero, Oaxaca). 79 MXS. Cl. tuber geophyte.

*Dioscorea daunea* Prain & Burkill, J. Proc. Asiat. Soc. Bengal 4: 450 (1908).
Indo-China to Pen. Malaysia. 41 MYA THA 42 MLY. Cl. tuber geophyte.

*Dioscorea davidsei* O.Téllez, Novon 7: 208 (1997).
Costa Rica to Panama. 80 COS PAN. Cl. tuber geophyte.

*Dioscorea de-mourae* Uline ex R.Knuth, Notizbl. Bot. Gart. Berlin-Dahlem 7: 199 (1917).
Brazil to NE. Argentina. 84 BZC BZE BZL BZN BZS 85 AGE. Cl. tuber geophyte.

*Dioscorea debilis* Uline ex R.Knuth, Notizbl. Bot. Gart. Berlin-Dahlem 7: 210 (1917).
Brazil (S. Bahia to Minas Gerais). 84 BZE BZL. Tuber geophyte.
*Dioscorea debilis* var. *sagittifolia* Uline ex R.Knuth, Notizbl. Bot. Gart. Berlin-Dahlem 7: 210 (1917).

*Dioscorea decaryana* H.Perrier, Notul. Syst. (Paris) 12: 205 (1946).
C. Madagascar. 29 MDG. Cl. tuber geophyte.

*Dioscorea decipiens* Hook.f., Fl. Brit. India 6: 293 (1892).
China (Yunnan) to Indo-China. 36 CHC 41 LAO MYA THA VIE. Cl. tuber geophyte.
*Dioscorea decipiens* var. *glabrescens* C.T.Ting & M.C.Chang, Acta Phytotax. Sin. 20: 206 (1982).

*Dioscorea decorticans* C.Presl, Reliq. Haenk. 1: 135 (1827). *Dioscorea amaranthoides* var. *decorticans* (C.Presl) Uline ex R.Knuth, Notizbl. Bot. Gart. Berlin-Dahlem 7: 215 (1917).
S. Venezuela to WC. Brazil. 82 VEN 83 PER 84 BZC. Cl. tuber geophyte.
*Dioscorea rubricaulis* Kunth, Enum. Pl. 5: 368 (1850).

*Dioscorea deflexa* Griseb., Vidensk. Meddel. Naturhist. Foren. Kjøbenhavn 1875: 157 (1875).
Brazil (Minas Gerais). 84 BZL. Cl. tuber geophyte.

*Dioscorea delavayi* Franch., Rev. Hort. 1896: 541 (1896). *Dioscorea kamoonensis* var. *delavayi* (Franch.) Prain & Burkill, Ann. Roy. Bot. Gard. (Calcutta) 14(1): 148 (1936).
SC. China. 36 CHC. Cl. tuber geophyte.
*Dioscorea kamoonensis* var. *henryi* Prain & Burkill, J. Proc. Asiat. Soc. Bengal 10: 22 (1914). *Dioscorea burkillii* R.Knuth in H.G.A.Engler (ed.), Pflanzenr., IV, 43: 143 (1924). *Dioscorea henryi* (Prain & Burkill) C.T.Ting, Acta Phytotax. Sin. 20: 208 (1982), nom. illeg.
*Dioscorea engleriana* R.Knuth in H.G.A.Engler (ed.), Pflanzenr., IV, 43: 140 (1924).
*Dioscorea rotundifoliolata* R.Knuth in H.G.A. Engler (ed.), Pflanzenr., IV, 43: 142 (1924).

*Dioscorea delicata* R.Knuth in H.G.A.Engler (ed.), Pflanzenr., IV, 43: 61 (1924).
SE. Brazil. 84 BZL. Cl. tuber geophyte.

*Dioscorea deltoidea* Wall. ex Griseb. in C.F.P.von Martius & auct. suc. (eds.), Fl. Bras. 3(1): 43 (1842).
Himalaya to SC. China. 36 CHC CHT 40 ASS BAN EHM NEP WHM 41 MYA THA VIE. Cl. rhizome geophyte.
*Tamus nepalensis* Jacquem. ex Prain & Burkill, J. Proc. Asiat. Soc. Bengal 10: 16 (1914).
*Dioscorea deltoidea* var. *orbiculata* Prain & Burkill, Ann. Roy. Bot. Gard. (Calcutta) 14: 25 (1936).
*Dioscorea nepalensis* Sweet ex Bernardi, Candollea 18: 258 (1963).

*Dioscorea dendrotricha* Uline in H.G.A.Engler & K.A.E.Prantl, Nat. Pflanzenfam., Nachtr. 1: 84 (1897).
N. Brazil. 84 BZN. Cl. tuber geophyte.

*Dioscorea densiflora* Hemsl., Biol. Cent.-Amer., Bot. 3: 356 (1884).
Mexico to C. America. 79 MXC MXE MXG MXS MXT 80 BLZ ELS GUA HON. Cl. tuber geophyte.

*Dioscorea depauperata* Prain & Burkill, Bull. Misc. Inform. Kew 1933: 245 (1933).
Indo-China. 41 LAO THA VIE. Cl. tuber geophyte.

*Dioscorea diamantinensis* R.Knuth in H.G.A.Engler (ed.), Pflanzenr., IV, 43: 231 (1924).
Brazil (Minas Gerais). 84 BZL. Cl. tuber geophyte.
    *Dioscorea heptaneura* f. *tenuicaulis* Uline ex R.Knuth, Notizbl. Bot. Gart. Berlin-Dahlem 7: 209 (1917).

*Dioscorea dicranandra* Donn.Sm., Bot. Gaz. 19: 13 (1894). *Dioscorea albicaulis* Uline, Bot. Jahrb. Syst. 22: 425 (1896), nom. illeg.
Guatemala. 80 GUA. Cl. tuber geophyte.

*Dioscorea dielsii* R.Knuth, Biblioth. Bot. 29(116): 69 (1937).
Ecuador. 83 ECU. Cl. tuber geophyte.

*Dioscorea dissimulans* Prain & Burkill, Bull. Misc. Inform. Kew 1933: 241 (1933).
S. Vietnam. 41 VIE. Cl. tuber geophyte.

*Dioscorea divaricata* Blanco, Fl. Filip.: 797 (1837).
Philippines. 42 PHI. Cl. tuber geophyte.
    *Dioscorea soror* Prain & Burkill, Leafl. Philipp. Bot. 5: 1598 (1913).
    *Dioscorea foxworthyi* Prain & Burkill, J. Proc. Asiat. Soc. Bengal 10: 34 (1914).
    *Dioscorea soror* var. *glauca* Prain & Burkill, J. Proc. Asiat. Soc. Bengal 10: 34 (1914).
    *Dioscorea soror* var. *vera* Prain & Burkill, J. Proc. Asiat. Soc. Bengal 10: 34 (1914).
    *Dioscorea oxyphylla* R.Knuth in H.G.A.Engler (ed.), Pflanzenr., IV, 43: 269 (1924).

*Dioscorea diversifolia* Griseb. in C.F.P.von Martius & auct. suc. (eds.), Fl. Bras. 3(1): 41 (1842).
S. Brazil. 84 BZS. Cl. tuber geophyte.

*Dioscorea dodecaneura* Vell., Fl. Flumin. 10: t. 123 (1831).
S. Trop. America. 82 FRG GUY SUR VEN 83 BOL ECU PER 84 BZC BZE BZL BZN BZS 85 AGE AGW PAR. Cl. tuber geophyte.
    *Dioscorea dodecandra* Steud., Nomencl. Bot., ed. 2, 1: 511 (1840), orth. var.
    *Dioscorea hebantha* Mart. ex Griseb. in C.F.P.von Martius & auct. suc. (eds.), Fl. Bras. 3(1): 39 (1842).
    *Dioscorea discolor* Kunth, Enum. Pl. 5: 334 (1850).
    *Dioscorea illustrata* W.Bull, Cat. 1873: 6 (1873).
    *Dioscorea vittata* W.Bull ex Baker, Bot. Mag. 105: t. 6409 (1879).
    *Dioscorea kita* Queva, Mém. Soc. Natl. Sci. Agric. Arts Lille, IV, 20: 326 (1894).
    *Dioscorea racemosa* Rusby, Bull. New York Bot. Gard. 4: 459 (1907), nom. illeg. *Dioscorea bangii* R.Knuth in H.G.A.Engler (ed.), Pflanzenr., IV, 43: 324 (1924).
    *Dioscorea dodecaneura* var. *maronensis* Uline ex R.Knuth, Notizbl. Bot. Gart. Berlin-Dahlem 7: 218 (1917).
    *Dioscorea dodecaneura* var. *villosa* R.Knuth in H.G.A.Engler (ed.), Pflanzenr., IV, 43: 250 (1924).
    *Dioscorea huallagensis* R.Knuth, Repert. Spec. Nov. Regni Veg. 29: 95 (1931).

*Dioscorea dregeana* (Kunth) T.Durand & Schinz, Consp. Fl. Afric. 5: 274 (1894).
S. Mozambique to S. Africa. 26 MOZ 27 CPP NAT SWZ. Cl. tuber geophyte.
    *Helmia dregeana* Kunth, Enum. Pl. 5: 437 (1850).

*Dioscorea duchassaingii* R.Knuth in H.G.A.Engler (ed.), Pflanzenr., IV, 43: 67 (1924).
Lesser Antilles. 81 LEE WIN. Cl. tuber geophyte.

*Dioscorea dugesii* C.B.Rob., Proc. Amer. Acad. Arts 29: 330 (1894).
Mexico. 79 MXE MXN MXS. Cl. tuber geophyte.
    *Dioscorea violacea* Uline, Bot. Jahrb. Syst. 22: 423 (1896).

*Dioscorea dumetorum* (Kunth) Pax in H.G.A.Engler & K.A.E.Prantl, Nat. Pflanzenfam. 2(5): 133 (1887).
Trop. & S. Africa. 22 BEN BKN GHA GUI IVO NGA SEN SIE TOG 23 BUR CAF CMN CON EQG GAB GGI RWA ZAI 24 CHA ETH SUD 25 KEN TAN UGA 26 ANG MLW MOZ ZAM ZIM 27 TVL (81) jam. Cl. tuber geophyte.
    *Dioscorea triphylla* Schimp. ex A.Rich., Tent. Fl. Abyss. 2: 316 (1850), nom. illeg.
    *Helmia dumetorum* Kunth, Enum. Pl. 5: 436 (1850).
    *Dioscorea triphylla* var. *dumetorum* (Kunth) R.Knuth in H.G.A.Engler (ed.), Pflanzenr., IV, 43: 132 (1924).
    *Dioscorea buchholziana* Engl., Bot. Jahrb. Syst. 7: 333 (1886).
    *Dioscorea triphylla* var. *tomentosa* Rendle in W.P.Hiern, Cat. Afr. Pl. 2(1): 40 (1899).
    *Dioscorea triphylla* var. *abyssinica* R.Knuth in H.G.A.Engler (ed.), Pflanzenr., IV, 43: 136 (1924).
    *Dioscorea triphylla* var. *rotundata* R.Knuth in H.G.A.Engler (ed.), Pflanzenr., IV, 43: 136 (1924).
    *Dioscorea triphylla* var. *rotundata* R.Knuth in H.G.A.Engler (ed.), Pflanzenr., IV, 43: 136 (1924).

*Dioscorea dumetosa* Uline ex R.Knuth, Notizbl. Bot. Gart. Berlin-Dahlem 7: 209 (1917).
Brazil (Mato Grosso). 84 BZC. Cl. tuber geophyte.

*Dioscorea ekmanii* R.Knuth, Notizbl. Bot. Gart. Berlin-Dahlem 7: 191 (1917).
WC. Cuba. 81 CUB. Cl. tuber geophyte.

*Dioscorea elegans* Ridl. ex Prain & Burkill, Bull. Misc. Inform. Kew 1925: 65 (1925).
SE. New Guinea. 43 NWG. Cl. tuber geophyte.
    *Dioscorea papuana* Ridl., Trans. Linn. Soc. London, Bot. 9: 227 (1916), nom. illeg.

*Dioscorea elephantipes* (L'Hér.) Engl. in H.G.A.Engler & C.G.O.Drude, Veg. Erde 9(3; 2): 367 (1908).
Cape Prov. 27 CPP. Tuber geophyte.
    *Tamus elephantipes* L'Hér., Sert. Angl.: 29 (1789).
    *Testudinaria elephantipes* (L'Hér.) Burch., Trav. S. Africa 2: 147 (1824). *Dioscorea elephantopus* Spreng., Syst. Veg. 4(2): 143 (1827), nom. illeg.
    *Rhizemys elephantipes* (L'Hér.) Raf., Fl. Tellur. 4: 26 (1838). *Dioscorea testudinaria* R.Knuth in H.G.A.Engler (ed.), Pflanzenr., IV, 43: 9 (1924), nom. illeg.
    *Testudinaria montana* Burch., Trav. S. Africa 2: 148 (1824). *Dioscorea montana* (Burch.) Spreng., Syst. Veg. 4(2): 143 (1827). *Rhizemys montana* (Burch.) Raf., Fl. Tellur. 4: 26 (1838). *Testudinaria elephantipes* f. *montana* (Burch.) G.D.Rowley, Natl. Cact. Succ. J. 28: 6 (1973).

*Dioscorea entomophila* Hauman, Anales Mus. Nac. Hist. Nat. Buenos Aires 27: 475 (1916).
Argentina (Tucumán). 85 AGW. Cl. tuber geophyte.
*Dioscorea entomophila* var. *tomentosa* Hauman, Anales Mus. Nac. Hist. Nat. Buenos Aires 27: 479 (1915).

*Dioscorea epistephioides* Taub., Bot. Jahrb. Syst. 21: 425 (1896).
WC. Brazil. 84 BZC. Cl. tuber geophyte.

*Dioscorea escuintlensis* Matuda, Bol. Soc. Bot. México 21: 4 (1957).
Guatemala. 80 GUA. Cl. tuber geophyte.

*Dioscorea esculenta* (Lour.) Burkill, Gard. Bull. Straits Settlem. 1: 396 (1917).
Trop. & Subtrop. Asia, cultivated elsewere. (22) ivo (29) mdg sey (36) chc CHH CHS (38) nns tai 40 BAN EHM NEP WHM 41 MYA THA VIE 42 BOR JAW LSI MLY MOL PHI SUL SUM 43 BIS NWG (60) fij nue ton (61) pit sci (62) crl mrn (81) lee win. Cl. tuber geophyte.
*Oncus esculentus* Lour., Fl. Cochinch.: 194 (1790).
*Oncorhiza esculenta* (Lour.) Pers., Syn. Pl. 1: 374 (1805).
*Dioscorea aculeata* Roxb., Fl. Ind. ed. 1832 3: 800 (1832), nom. illeg.
*Dioscorea fasciculata* Roxb., Fl. Ind. ed. 1832, 3: 801 (1832). *Dioscorea esculenta* var. *fasciculata* (Roxb.) Prain & Burkill, J. Proc. Asiat. Soc. Bengal 10: 19 (1914).
*Dioscorea papillaris* Blanco, Fl. Filip.: 801 (1837).
*Dioscorea tugui* Blanco, Fl. Filip.: 800 (1837).
*Dioscorea tiliifolia* Kunth, Enum. Pl. 5: 401 (1850). *Dioscorea esculenta* var. *tiliifolia* (Kunth) Fosberg & Sachet, Micronesica 20: 135 (1978).
*Dioscorea papuana* Warb., Bot. Jahrb. Syst. 13: 273 (1891).
*Dioscorea spinosa* Roxb. ex Hook.f., Fl. Brit. India 6: 291 (1892).
*Dioscorea aculeata* var. *spinosa* Roxb. ex Prain & Burkill, J. Proc. Asiat. Soc. Bengal 10: 20 (1914). *Dioscorea esculenta* var. *spinosa* (Roxb. ex Prain & Burkill) R.Knuth in H.G.A.Engler (ed.), Pflanzenr., IV, 43: 189 (1924).
*Dioscorea esculenta* var. *fulvidotomentosa* R.Knuth in H.G.A.Engler (ed.), Pflanzenr., IV, 43: 190 (1924).

*Dioscorea esquirolii* Prain & Burkill, Bull. Misc. Inform. Kew 1931: 426 (1931).
China (Yunnan. S. Guizhou, Guangxi). 36 CHC CHS. Cl. tuber geophyte.

*Dioscorea exalata* C.T.Ting & M.C.Chang, Acta Phytotax. Sin. 20: 208 (1982).
S. China to Indo-China. 36 CHC CHS 41 THA VIE. Cl. tuber geophyte.

*Dioscorea fandra* H.Perrier, Mém. Soc. Linn. Normandie, Bot. 1(1): 27 (1928).
SW. Madagascar. 29 MDG. Cl. tuber geophyte.

*Dioscorea fasciculocongesta* (Sosa & B.G.Schub.) O.Téllez, Contr. Univ. Michigan Herb. 21: 310 (1997).
Mexico to C. America. 79 MXC MXG MXS 80 PAN. Cl. tuber geophyte.
*Dioscorea spiculiflora* var. *fasciculocongesta* Sosa & B.G.Schub., Biótica 11: 187 (1986).

*Dioscorea fastigiata* Gay, Fl. Chil. 6: 54 (1854).
N. Chile (to Coquimbo). 85 CLC CLN. Tuber geophyte.

*Dioscorea axilliflora* Phil., Anales Univ. Chile 93: 11 (1896).
*Dioscorea gayi* Phil., Anales Univ. Chile 93: 9 (1896).
*Dioscorea geissei* Phil., Anales Univ. Chile 93: 8 (1896).
*Dioscorea paupera* Phil., Anales Univ. Chile 93: 12 (1896).
*Dioscorea thinophila* Phil., Anales Univ. Chile 93: 10 (1896).
*Dioscorea cylindrostachya* I.M.Johnst., Contr. Gray Herb. 85: 25 (1929).

*Dioscorea fendleri* R.Knuth in H.G.A.Engler (ed.), Pflanzenr., IV, 43: 65 (1924).
Venezuela. 82 VEN. Cl. tuber geophyte.

*Dioscorea ferreyrae* Ayala, Diosc. Peru: 27 (1998).
Peru. 83 PER. Cl. tuber geophyte.

*Dioscorea filicaulis* Prain & Burkill, Bull. Misc. Inform. Kew 1933: 242 (1933).
Thailand. 41 THA. Cl. tuber geophyte.

*Dioscorea filiformis* Blume, Enum. Pl. Javae 1: 22 (1827).
Thailand, Malesia. 41 THA 42 BOR JAW LSI MLY MOL PHI SUL SUM. Cl. tuber geophyte.
*Dioscorea repanda* Blume, Enum. Pl. Javae 1: 22 (1827).
*Dioscorea myriantha* Kunth, Enum. Pl. 5: 382 (1850).
*Dioscorea gibbiflora* Hook.f., Fl. Brit. India 6: 294 (1892).
*Dioscorea koordersii* R.Knuth in H.G.A.Engler (ed.), Pflanzenr., IV, 43: 291 (1924).

*Dioscorea fimbriata* Jum. & H.Perrier, Ann. Inst. Bot.-Géol. Colon. Marseille, II, 8: 291 (1910).
Madagascar. 29 MDG. Cl. tuber geophyte.

*Dioscorea flabellifolia* Prain & Burkill, Leafl. Philipp. Bot. 5: 1593 (1913).
Borneo to Caroline Is. 42 BOR PHI 62 CRL. Cl. tuber geophyte.
*Dioscorea ledermannii* R.Knuth in H.G.A.Engler (ed.), Pflanzenr., IV, 43: 188 (1924).
*Dioscorea bullata* Prain & Burkill, Bull. Misc. Inform. Kew 1925: 60 (1925).

*Dioscorea flaccida* R.Knuth, Repert. Spec. Nov. Regni Veg. 28: 85 (1930).
Colombia. 83 CLM. Cl. tuber geophyte.

*Dioscorea floribunda* M.Martens & Galeotti, Bull. Acad. Roy. Sci. Bruxelles 9(2): 391 (1842).
S. Mexico to C. America. 79 MXS MXT 80 BLZ ELS GUA HON. Cl. tuber geophyte.
*Dioscorea barclayi* R.Knuth in H.G.A.Engler (ed.), Pflanzenr., IV, 43: 169 (1924).

*Dioscorea floridana* Bartlett, Bull. Bur. Pl. Industr. U.S.D.A. 189: 18 (1910). *Dioscorea villosa* subsp. *floridana* (Bartlett) R.Knuth in H.G.A.Engler (ed.), Pflanzenr., IV, 43: 173 (1924). *Dioscorea villosa* var. *floridana* (Bartlett) H.E.Ahles, J. Elisha Mitchell Sci. Soc. 80: 172 (1964).
SE. U.S.A. 78 FLA GEO SCA. Cl. rhizome geophyte.

*Dioscorea fodinarum* Kunth, Enum. Pl. 5: 405 (1850).
Brazil to Peru. 82 VEN 83 PER 84 BZC BZE BZL. Cl. tuber geophyte.
*Dioscorea multiflora* var. *grandifolia* Griseb. in C.F.P.von Martius & auct. suc. (eds.), Fl. Bras. 3(1): 35 (1842). *Dioscorea venosa* Uline ex R.Knuth, Notizbl. Bot. Gart. Berlin-Dahlem 7: 190 (1917).
*Dioscorea effusa* Griseb., Vidensk. Meddel. Naturhist. Foren. Kjøbenhavn 1875: 161 (1875).

*Dioscorea venosa* var. *effusa* Uline, Notizbl. Bot. Gart. Berlin-Dahlem 7: 191 (1917).
*Dioscorea venosa* var. *fodinarum* Uline, Notizbl. Bot. Gart. Berlin-Dahlem 7: 191 (1917).

**Dioscorea fordii** Prain & Burkill, J. Proc. Asiat. Soc. Bengal 4: 450 (1908).
SE. China, Hainan. 36 CHH CHS. Cl. tuber geophyte.
*Dioscorea hainanensis* Prain & Burkill, Bull. Misc. Inform. Kew 1936: 494 (1936).

**Dioscorea formosana** R.Knuth in H.G.A.Engler (ed.), Pflanzenr., IV, 43: 268 (1924).
Taiwan. 38 TAI. Cl. tuber geophyte.

**Dioscorea fractiflexa** R.Knuth in H.G.A.Engler (ed.), Pflanzenr., IV, 43: 235 (1924).
S. Brazil. 84 BZS. Cl. tuber geophyte.

**Dioscorea fuliginosa** R.Knuth, Repert. Spec. Nov. Regni Veg. 21: 78 (1925).
Bolivia. 83 BOL. Cl. tuber geophyte.

**Dioscorea furcata** Griseb. in C.F.P.von Martius & auct. suc. (eds.), Fl. Bras. 3(1): 45 (1842). *Helmia furcata* (Griseb.) Kunth, Enum. Pl. 5: 419 (1850).
S. Brazil to Uruguay. 84 BZS 85 AGE URU. Cl. tuber geophyte.
*Dioscorea fracta* Griseb. in C.F.P.von Martius & auct. suc. (eds.), Fl. Bras. 3(1): 44 (1842). *Helmia fracta* (Griseb.) Kunth, Enum. Pl. 5: 420 (1850).

**Dioscorea futschauensis** Uline ex R.Knuth in H.G.A.Engler (ed.), Pflanzenr., IV, 43: 264 (1924).
SE. China. 36 CHS. Cl. rhizome geophyte.

**Dioscorea galeottiana** Kunth, Enum. Pl. 5: 409 (1850).
*Dioscorea convolvulacea* var. *galeottiana* (Kunth) Uline, Bot. Jahrb. Syst. 22: 427 (1896).
Mexico. 79 MXC MXG MXS MXT. Cl. tuber geophyte.
*Dioscorea grandiflora* M.Martens & Galeotti, Bull. Acad. Roy. Sci. Bruxelles 9(2): 392 (1842), nom. illeg.
*Dioscorea lobata* Uline, Bot. Jahrb. Syst. 22: 427 (1896).
*Dioscorea convolvulacea* var. *viridis* Uline, Bot. Jahrb. Syst. 22: 427 99 (1897).
*Dioscorea lobata* var. *lasiophylla* Uline ex R.Knuth, Notizbl. Bot. Gart. Berlin-Dahlem 7: 194 (1917).

**Dioscorea galiiflora** R.Knuth, Notizbl. Bot. Gart. Berlin-Dahlem 7: 538 (1921).
WC. Brazil. 84 BZC. Cl. tuber geophyte.

**Dioscorea gallegosi** Matuda, Anales Inst. Biol. Univ. Nac. México 24: 288 (1954).
C. Mexico. 79 MXC. Cl. tuber geophyte.

**Dioscorea garrettii** Prain & Burkill, Bull. Misc. Inform. Kew 1936: 493 (1936).
China (Yunnan) to Thailand. 41 THA. Cl. tuber geophyte.

**Dioscorea gaumeri** R.Knuth, Notizbl. Bot. Gart. Berlin-Dahlem 7: 199 (1917).
SE. Mexico to Belize. 79 MXT 80 BLZ. Cl. tuber geophyte.
*Higinbothamia synandra* Uline, Publ. Field Columbian Mus., Bot. Ser. 1: 415 (1899). *Dioscorea synandra* (Uline) Standl., Publ. Field Columbian Mus., Bot. Ser. 3: 231 (1930), nom. illeg. *Dioscorea dodecasemina* Caddick & Wilkin, Taxon 51: 112 (2002).

**Dioscorea gentryi** O.Téllez, Brittonia 48: 100 (1996).
Peru. 83 PER. Cl. tuber geophyte.

**Dioscorea gillettii** Milne-Redh., Kew Bull. 17: 177 (1963).
S. Ethiopia to Kenya. 24 ETH 25 KEN. Tuber geophyte.

**Dioscorea glabra** Roxb., Fl. Ind. ed. 1832, 3: 804 (1832).
Himalaya to Malaya. 36 CHC CHS 40 ASS BAN EHM IND NEP WHM 41 CBD LAO MYA THA VIE 42 MLY. Cl. tuber geophyte.
*Dioscorea glabra* var. *hastifolia* Prain & Burkill, J. Proc. Asiat. Soc. Bengal 10: 37 (1914).
*Dioscorea hongkongensis* Uline ex R.Knuth in H.G.A.Engler (ed.), Pflanzenr., IV, 43: 288 (1924).
*Dioscorea siamensis* R.Knuth in H.G.A.Engler (ed.), Pflanzenr., IV, 43: 281 (1924).
*Dioscorea glabra* var. *vera* Prain & Burkill, Ann. Roy. Bot. Gard. (Calcutta) 14: 357 (1936).

**Dioscorea glandulosa** (Griseb.) Klotzsch ex Kunth, Enum. Pl. 5: 352 (1850).
W. South America to N. Argentina, Brazil. 83 BOL CLM PER 84 BZC BZE BZL BZN BZS 85 AGW PAR. Cl. tuber geophyte.
\*Dioscorea piperifolia* var. *glandulosa* Griseb. in C.F.P.von Martius & auct. suc. (eds.), Fl. Bras. 3(1): 27 (1842).

var. **calcensis** (R.Knuth) Ayala, Diosc. Peru: 29 (1998).
Peru. 83 PER. Cl. tuber geophyte.
\*Dioscorea calcensis* R.Knuth, Repert. Spec. Nov. Regni Veg. 30: 159 (1932).

var. **glandulosa**
W. South America to N. Argentina, Brazil. 83 BOL CLM PER 84 BZC BZE BZL BZN BZS 85 AGW PAR. Cl. tuber geophyte.
*Dioscorea undecimnervis* Vell., Fl. Flumin. 10: t. 120 (1831), provisional synonym.
*Dioscorea piperifolia* var. *legitima* Griseb. in C.F.P.von Martius & auct. suc. (eds.), Fl. Bras. 3(1): 27 (1842).
*Dioscorea piperifolia* var. *triangularis* Griseb. in C.F.P.von Martius & auct. suc. (eds.), Fl. Bras. 3(1): 27 (1842). *Dioscorea triangularis* (Griseb.) R.Knuth in H.G.A.Engler (ed.), Pflanzenr., IV, 43: 57 (1924), nom. illeg.
*Dioscorea sororia* Kunth, Enum. Pl. 5: 353 (1850).
*Dioscorea multispicata* R.Knuth, Meded. Rijks-Herb. 29: 55 (1916).
*Dioscorea paranensis* R.Knuth, Notizbl. Bot. Gart. Berlin-Dahlem 7: 195 (1917).
*Dioscorea sulcata* R.Knuth, Notizbl. Bot. Gart. Berlin-Dahlem 7: 195 (1917).
*Dioscorea recurva* Rusby, Descr. S. Amer. Pl.: 4 (1920).

**Dioscorea glomerulata** Hauman, Anales Mus. Nac. Hist. Nat. Buenos Aires 27: 455 (1916).
Bolivia to NW. Argentina. 83 BOL 85 AGW. Cl. tuber geophyte.
*Dioscorea boliviensis* R.Knuth, Notizbl. Bot. Gart. Berlin-Dahlem 7: 188 (1917).
*Dioscorea lorentzii* Uline ex R.Knuth, Notizbl. Bot. Gart. Berlin-Dahlem 7: 187 (1917).
*Dioscorea lorentzii* var. *mandonii* Uline ex R.Knuth, Notizbl. Bot. Gart. Berlin-Dahlem 7: 187 (1917).
*Dioscorea glomerulata* var. *mandonii* (Uline ex R.Knuth) R.Knuth in H.G.A.Engler (ed.), Pflanzenr., IV, 43: 59 (1924).

*Dioscorea nodosa* R.Knuth, Notizbl. Bot. Gart. Berlin-Dahlem 7: 187 (1917).
*Dioscorea burroyacensis* R.Knuth, Repert. Spec. Nov. Regni Veg. 40: 220 (1936).

**Dioscorea gomez-pompae** O.Téllez, Contr. Univ. Michigan Herb. 21: 309 (1997).
SE. Mexico to Guatemala. 79 MXT 80 GUA. Cl. tuber geophyte.
*\*Dioscorea spiculiflora* var. *chiapasana* Gómez Pompa, Ciencia (Mexico) 18: 242 (1959).

**Dioscorea gracilicaulis** R.Knuth, Notizbl. Bot. Gart. Berlin-Dahlem 7: 216 (1917).
Brazil (Goiás). 84 BZC. Cl. tuber geophyte.

**Dioscorea gracilipes** Prain & Burkill, Bull. Misc. Inform. Kew 1925: 63 (1925).
Pen. Thailand. 41 THA. Cl. rhizome geophyte.

**Dioscorea gracilis** Hook. ex Poepp., Fragm. Syn. Pl.: 12 (1833). *Dioscorea humifusa* var. *gracilis* (Hook. ex Poepp.) L.E.Navas, Anales Acad. Chilena Ci. Nat. 1968: 48 (1968).
C. Chile. 85 CLC. Cl. tuber geophyte.

**Dioscorea gracillima** Miq., Ann. Mus. Bot. Lugduno-Batavi 3: 160 (1867).
C. & S. Japan, S. China. 36 CHC CHS 38 JAP. Cl. rhizome geophyte.

**Dioscorea grandiflora** Mart. ex Griseb. in C.F.P.von Martius & auct. suc. (eds.), Fl. Bras. 3(1): 28 (1842).
Brazil to Paraguay. 84 BZC BZL 85 PAR. Cl. tuber geophyte.

**Dioscorea grandis** R.Knuth, Notizbl. Bot. Gart. Berlin-Dahlem 7: 194 (1917).
Peru. 83 PER. Cl. tuber geophyte.

**Dioscorea grata** Prain & Burkill, J. Proc. Asiat. Soc. Bengal 10: 35 (1914).
Philippines to New Guinea. 42 PHI 43 NWG. Cl. tuber geophyte.

**Dioscorea gribinguiensis** Baudon, Ann. Inst. Bot.-Géol. Colon. Marseille, III, 1: 241 (1913).
Central African Rep. 23 CAF. Cl. tuber geophyte.

**Dioscorea grisebachii** Kunth, Enum. Pl. 5: 853 (1850).
Brazil. 84 BZC BZE BZL BZN BZS. Cl. tuber geophyte.
*\*Dioscorea filiformis* Griseb. in C.F.P.von Martius & auct. suc. (eds.), Fl. Bras. 3(1): 44 (1842), nom. illeg. *Hyperocarpa filiformis* G.M.Barroso, E.F.Guim. & Sucre, Sellowia 25: 20 (1974).
*Dioscorea kunthiana* Uline ex R.Knuth, Notizbl. Bot. Gart. Berlin-Dahlem 7: 211 (1917).

**Dioscorea guerrerensis** R.Knuth, Repert. Spec. Nov. Regni Veg. 40: 221 (1936).
C. & SW. Mexico. 79 MXC MXS. Cl. tuber geophyte.

**Dioscorea guianensis** R.Knuth, Notizbl. Bot. Gart. Berlin-Dahlem 7: 191 (1917).
Guyana. 82 GUY. Cl. tuber geophyte.

**Dioscorea haenkeana** C.Presl, Reliq. Haenk. 1: 135 (1827).
Panama to Peru. 80 PAN 83 PER. Cl. tuber geophyte.

**Dioscorea hamiltonii** Hook.f., Fl. Brit. India 6: 294 (1892).
Nepal to Indo-China. 40 ASS BAN EHM IND NEP 41 MYA THA VIE. Cl. tuber geophyte.

**Dioscorea hassleriana** Chodat, Bull. Herb. Boissier, II, 3: 1111 (1903). *Dioscorea hastata* var. *hassleriana* (Chodat) Uline ex R.Knuth, Notizbl. Bot. Gart. Berlin-Dahlem 7: 213 (1917).
Brazil to NW. Argentina. 83 BOL 84 BZC BZE BZL BZS 85 AGW PAR. Cl. tuber geophyte.
*Dioscorea hastata* Vell., Fl. Flumin. 10: t. 126 (1831), nom. illeg.
*Dioscorea hassleriana* var. *triloba* Chodat & Hassl., Bull. Herb. Boissier, II, 3: 111 (1903).
*Dioscorea hastata* var. *balansae* Uline ex R.Knuth, Notizbl. Bot. Gart. Berlin-Dahlem 7: 213 (1917).
*Dioscorea hastata* var. *mattogrossensis* Uline ex R.Knuth, Notizbl. Bot. Gart. Berlin-Dahlem 7: 213 (1917).
*Dioscorea platystemon* Hauman, Anales Mus. Nac. Hist. Nat. Buenos Aires 27: 493 (1918).

**Dioscorea hastata** Mill., Gard. Dict. ed. 8: 2 (1768).
French Guiana. 82 FRG. Cl. tuber geophyte.

**Dioscorea hastatissima** Rusby, Descr. S. Amer. Pl.: 7 (1920).
Colombia. 83 CLM. Cl. tuber geophyte.

**Dioscorea hastifolia** Nees in J.G.C.Lehmann, Pl. Preiss. 2: 33 (1846).
WSW. Western Australia. 50 WAU. Cl. tuber geophyte.

**Dioscorea hastiformis** R.Knuth, Notizbl. Bot. Gart. Berlin-Dahlem 7: 197 (1917).
S. Bolivia. 83 BOL. Cl. tuber geophyte.

**Dioscorea haumanii** Xifreda, Bol. Soc. Argent. Bot. 20: 317 (1982).
Paraguay to NW. Argentina. 85 AGW PAR. Cl. tuber geophyte.

**Dioscorea havilandii** Prain & Burkill, J. Proc. Asiat. Soc. Bengal 10: 40 (1914).
Borneo, Sumatera (Belitung). 42 BOR SUM. Cl. tuber geophyte.

**Dioscorea hebridensis** R.Knuth, Repert. Spec. Nov. Regni Veg. 40: 224 (1936).
Vanuatu. 60 VAN. Cl. tuber geophyte.

**Dioscorea hemicrypta** Burkill, J. S. African Bot. 18: 187 (1952).
Cape Prov. 27 CPP. Caudex geophyte.

**Dioscorea hemsleyi** Prain & Burkill, J. Proc. Asiat. Soc. Bengal 4: 451 (1908).
C. China to Indo-China. 36 CHC CHS 41 CBD LAO MYA VIE. Cl. tuber geophyte.
*Dioscorea praecox* Prain & Burkill, J. Proc. Asiat. Soc. Bengal 4: 455 (1908).
*Dioscorea mairei* H.Lév., Repert. Spec. Nov. Regni Veg. 12: 288 (1913).

**Dioscorea heptaneura** Vell., Fl. Flumin. 10: t. 124 (1831).
SE. Brazil. 84 BZE? BZL. Cl. tuber geophyte.
*Dioscorea heptaneura* f. *latisinuata* Uline ex R.Knuth, Notizbl. Bot. Gart. Berlin-Dahlem 7: 209 (1917).

**Dioscorea herbert-smithii** Rusby, Descr. S. Amer. Pl.: 7 (1920).
Colombia. 83 CLM. Cl. tuber geophyte.

**Dioscorea herzogii** R.Knuth, Meded. Rijks-Herb. 29: 56 (1916).
Bolivia. 83 BOL. Cl. tuber geophyte.

**Dioscorea heteropoda** Baker, J. Bot. 20: 270 (1882).
C. Madagascar. 29 MDG. Tuber geophyte.

*Dioscorea hexagona* Baker, J. Bot. 20: 270 (1882).
C. Madagascar. 29 MDG. Tuber geophyte.

*Dioscorea hieronymi* Uline ex R.Knuth, Notizbl. Bot. Gart. Berlin-Dahlem 7: 197 (1917).
NW. Argentina. 85 AGW. Cl. tuber geophyte.
*Dioscorea megalantha* var. *subsessilis* Hauman, Anales Mus. Nac. Hist. Nat. Buenos Aires 27: 473 (1915).
*Dioscorea chiquiacensis* R.Knuth, Notizbl. Bot. Gart. Berlin-Dahlem 7: 196 (1917).
*Dioscorea tafiensis* R.Knuth, Repert. Spec. Nov. Regni Veg. 40: 222 (1936).

*Dioscorea hintonii* R.Knuth, Repert. Spec. Nov. Regni Veg. 40: 222 (1936).
C. & SW. Mexico. 79 MXC MXS. Cl. tuber geophyte.

*Dioscorea hirtiflora* Benth. in W.J.Hooker, Niger Fl.: 537 (1849).
Trop. Africa to Caprivi Strip. 22 BEN GHA GUI IVO LBR NGA SEN SIE TOG 23 CMN GAB ZAI 25 KEN TAN 26 ANG MLW MOZ ZAM ZIM 27 CPV. Cl. tuber geophyte.

subsp. *hirtiflora*
W. Trop. Africa to Angola. 22 BEN GHA GUI IVO LBR NGA SEN SIE TOG 23 CMN GAB ZAI 26 ANG. Cl. tuber geophyte.
*Dioscorea rubiginosa* Benth. in W.J.Hooker, Niger Fl.: 538 (1849).
*Dioscorea polyantha* Rendle in W.P.Hiern, Cat. Afr. Pl. 2: 37 (1899).
*Dioscorea anthropophagorum* var. *sylvestris* A.Chev., Veg. Ut. Afr. Trop. Franç. 8: 357 (1913).
*Dioscorea dusenii* Uline ex R.Knuth in H.G.A. Engler (ed.), Pflanzenr., IV, 43: 257 (1924).

subsp. *orientalis* Milne-Redh., Kew Bull. 26: 574 (1972).
Kenya to Mozambique. 25 KEN TAN 26 MLW MOZ. Cl. tuber geophyte.
*Dioscorea lindiensis* R.Knuth, Notizbl. Bot. Gart Berlin-Dahlem 12: 703 (1935).

subsp. *pedicellata* Milne-Redh., Kew Bull. 26: 573 (1972).
Uganda to S. Trop. Africa. 23 ZAI 25 TAN UGA 26 MLW MOZ ZAM ZIM 27 CPV. Cl. tuber geophyte.

*Dioscorea hispida* Dennst., Schlüssel Hortus Malab.: 33 (1818).
Trop. & Subtrop. Asia to N. Australia. 36 CHH CHS CHT 38 TAI 40 ASS BAN EHM IND NEP 41 AND MYA THA VIE 42 BOR JAW LSI MLY MOL PHI SUL SUM 43 BIS NWG 50 QLD. Cl. tuber geophyte.
*Dioscorea lunata* Roth, Nov. Pl. Sp.: 370 (1821).
*Dioscorea hirsuta* Blume, Enum. Pl. Javae 1: 21 (1827). *Helmia hirsuta* (Blume) Kunth, Enum. Pl. 5: 438 (1850).
*Dioscorea mollissima* Blume, Enum. Pl. Javae 1: 21 (1827). *Dioscorea triphylla* var. *mollissima* (Blume) Prain & Burkill, J. Proc. Asiat. Soc. Bengal 10(1): 26 (1914).
*Dioscorea daemona* Roxb., Fl. Ind. ed. 1832, 3: 805 (1832). *Helmia daemona* (Roxb.) Kunth, Enum. Pl. 5: 439 (1850). *Dioscorea triphylla* var. *daemona* (Roxb.) Prain & Burkill, J. Proc. Asiat. Soc. Bengal 10(1): 26 (1914). *Dioscorea hispida* var. *daemona* (Roxb.) Prain & Burkill, Ann. Roy. Bot. Gard. (Calcutta) 14: 192 (1936).

*Dioscorea daemona* var. *reticulata* Hook.f., Fl. Brit. India 6: 289 (1892). *Dioscorea triphylla* var. *reticulata* (Hook.f.) Prain & Burkill, J. Proc. Asiat. Soc. Bengal 10: 26 (1914). *Dioscorea hispida* var. *reticulata* (Hook.f.) Sanjappa in S.Karthikeyan & al., Fl. Ind. Enumerat. – Monocot.: 74 (1989).

*Dioscorea holmioidea* Maury, J. Bot. (Morot) 3: 267 (1889).
Venezuela (Amazonas). 82 VEN. Cl. tuber geophyte.

*Dioscorea hombuka* H.Perrier, Mém. Soc. Linn. Normandie, Bot. 1(1): 31 (1928).
Madagascar. 29 MDG. Cl. tuber geophyte.

*Dioscorea hondurensis* R.Knuth, Repert. Spec. Nov. Regni Veg. 38: 120 (1935).
SE. Mexico to C. America. 79 MXT 80 BLZ COS GUA HON NIC PAN. Cl. tuber geophyte.
*Dioscorea belizensis* Lundell, Contr. Univ. Michigan Herb. 6: 5 (1941).
*Dioscorea tabascana* Matuda, Bol. Soc. Bot. México 21: 1 (1957).

*Dioscorea humifusa* Poepp., Fragm. Syn. Pl.: 12 (1833).
C. Chile. 85 CLC. Cl. tuber geophyte.
*Dioscorea filipendula* Dombey ex Kunth, Enum. Pl. 5: 341 (1850).
*Dioscorea nigrescens* Phil., Anales Univ. Chile 93: 5 (1902).

*Dioscorea humilis* Bertero ex Colla, Mem. Reale Accad. Sci. Torino 39: 12 (1836). *Epipetrum humile* (Bertero ex Colla) Phil., Linnaea 33: 253 (1865). *Borderea humilis* (Bertero ex Colla) Pax in H.G.A.Engler & K.A.E.Prantl, Nat. Pflanzenfam. 2(5): 133 (1887).
C. Chile. 85 CLC. Tuber geophyte.
*Dioscorea pusilla* Hook., Hooker's Icon. Pl. 7: t. 678 (1844).

*Dioscorea hunzikeri* Xifreda, Darwiniana 26: 371 (1985).
Bolivia. 83 BOL. Cl. tuber geophyte.

*Dioscorea igualamontana* Matuda, Anales Inst. Biol. Univ. Nac. México 24: 56 (1953).
Mexico (Guerrero). 79 MXS. Cl. tuber geophyte.

*Dioscorea incayensis* R.Knuth, Repert. Spec. Nov. Regni Veg. 28: 85 (1930).
Peru. 83 PER. Cl. tuber geophyte.
*Dioscorea vargasii* Standl., Publ. Field Mus. Nat. Hist., Bot. Ser. 22: 134 (1940).

*Dioscorea inopinata* Prain & Burkill, Bull. Misc. Inform. Kew 1927: 245 (1927).
SW. Thailand. 41 THA. Cl. tuber geophyte.

*Dioscorea insignis* C.V.Morton & B.G.Schub., Proc. Biol. Soc. Wash. 84: 447 (1972).
Mexico (Guerrero). 79 MXS. Cl. tuber geophyte.

*Dioscorea intempestiva* Prain & Burkill, Bull. Misc. Inform. Kew 1933: 243 (1933).
Indo-China. 41 THA VIE. Cl. tuber geophyte.

*Dioscorea intermedia* Thwaites, Enum. Pl. Zeyl.: 326 (1864).
Sri Lanka. 40 SRL. Cl. tuber geophyte.

*Dioscorea ionophylla* Uline ex R.Knuth, Notizbl. Bot. Gart. Berlin-Dahlem 7: 207 (1917).
SC. Chile. 85 CLC. Cl. tuber geophyte.

*Dioscorea iquitosensis* R.Knuth, Repert. Spec. Nov. Regni Veg. 29: 94 (1931).
Peru. 83 PER. Cl. tuber geophyte.

*Dioscorea irupanensis* R.Knuth, Repert. Spec. Nov. Regni Veg. 30: 160 (1932).
Bolivia. 83 BOL. Cl. tuber geophyte.

*Dioscorea itapirensis* R.Knuth, Notizbl. Bot. Gart. Berlin-Dahlem 7: 190 (1917).
SE. Brazil. 84 BZL. Cl. tuber geophyte.

*Dioscorea itatiensis* R.Knuth, Notizbl. Bot. Gart. Berlin-Dahlem 7: 212 (1917).
Brazil (Rio de Janeiro). 84 BZL. Cl. tuber geophyte.

*Dioscorea jaliscana* S.Watson, Proc. Amer. Acad. Arts 22: 458 (1887).
C. & SW. Mexico. 79 MXC MXS. Cl. tuber geophyte.
*Dioscorea hirsuticaulis* C.B.Rob., Proc. Amer. Acad. Arts 29: 324 (1894).

*Dioscorea jamesonii* R.Knuth in H.G.A.Engler (ed.), Pflanzenr., IV, 43: 75 (1924).
Ecuador. 83 ECU. Cl. tuber geophyte.

*Dioscorea japonica* Thunb. in J.A.Murray, Syst. Veg. ed. 14: 889 (1784).
S. China to Assam, Temp. E. Asia. 36 CHC CHS 38 JAP KOR NNS oga TAI 40 ASS. Cl. tuber geophyte.

var. *japonica*
S. China, Temp. E. Asia. 36 CHC CHS 38 JAP KOR NNS oga TAI. Cl. tuber geophyte.
*Dioscorea goeringiana* Kunth, Enum. Pl. 5: 402 (1850).
*Dioscorea belophylloides* Prain & Burkill, J. Proc. Asiat. Soc. Bengal 4: 448 (1908).
*Dioscorea japonica* var. *vera* Prain & Burkill, J. Proc. Asiat. Soc. Bengal 10: 28 (1914).
*Dioscorea pseudojaponica* Hayata, Icon. Pl. Formosan. 10: 42 (1921). *Dioscorea japonica* var. *pseudojaponica* (Hayata) Yamam., J. Soc. Trop. Agric. 10: 182 (1938).
*Dioscorea fauriei* R.Knuth in H.G.A.Engler (ed.), Pflanzenr., IV, 43: 263 (1924).
*Dioscorea kelungensis* R.Knuth in H.G.A.Engler (ed.), Pflanzenr., IV, 43: 263 (1924), nom. illeg. *Dioscorea neglecta* R.Knuth in H.G.A.Engler (ed.), Pflanzenr., IV, 43: 355 (1924). *Dioscorea japonica* var. *kelungensis* Prain & Burkill, Bull. Misc. Inform. Kew 1926: 119 (1926).
*Dioscorea kiangsiensis* R.Knuth, Repert. Spec. Nov. Regni Veg. 21: 80 (1925).

var. *nagarum* Prain & Burkill, Ann. Roy. Bot. Gard. (Calcutta) 14: 259 (1936).
Assam. 40 ASS. Cl. tuber geophyte.

var. *oldhamii* R.Knuth in H.G.A.Engler (ed.), Pflanzenr., IV, 43: 263 (1924).
SE. China, Taiwan. 36 CHS 38 TAI. Cl. tuber geophyte.

var. *pilifera* C.T.Ting & M.C.Chang, Acta Phytotax. Sin. 20: 206 (1982).
S. China. 36 CHC CHS. Cl. tuber geophyte.

*Dioscorea javariensis* Ayala, Diosc. Peru: 33 (1998).
Peru. 83 PER. Cl. tuber geophyte.

*Dioscorea juxtlahuacensis* (O.Téllez & Dávila) Caddick & Wilkin, Taxon 51: 112 (2002).
SW. Mexico. 79 MXS. Cl. tuber geophyte.
*Nanarepenta juxtlahuacensis* O.Téllez & Dávila, Novon 8: 210 (1998).

*Dioscorea kalkapershadii* Prain & Burkill, J. Proc. Asiat. Soc. Bengal 10: 24 (1914).
India. 40 IND. Cl. tuber geophyte.

*Dioscorea kamoonensis* Kunth, Enum. Pl. 5: 395 (1850).
Himalaya to S. China and Indo-China. 36 CHC CHS CHT 40 ASS BAN EHM NEP WHM 41 LAO MYA THA VIE. Cl. tuber geophyte.
*Dioscorea fargesii* Franch., Rev. Hort. 1896: 540 (1896).
*Dioscorea kamoonensis* var. *fargesii* (Franch.) Prain & Burkill, J. Proc. Asiat. Soc. Bengal 10: 21 (1914).
*Dioscorea kamoonensis* var. *straminea* Prain & Burkill, J. Proc. Asiat. Soc. Bengal 10: 21 (1914).
*Dioscorea firma* R.Knuth in H.G.A.Engler (ed.), Pflanzenr., IV, 43: 141 (1924).
*Dioscorea mairei* R.Knuth in H.G.A.Engler (ed.), Pflanzenr., IV, 43: 144 (1924), nom. illeg.
*Dioscorea mengtzeana* R.Knuth in H.G.A.Engler (ed.), Pflanzenr., IV, 43: 142 (1924).
*Dioscorea subfusca* R.Knuth in H.G.A.Engler (ed.), Pflanzenr., IV, 43: 143 (1924).
*Dioscorea bonatiana* Prain & Burkill, Bull. Misc. Inform. Kew 1925: 61 (1925).
*Dioscorea kamoonensis* var. *praecox* Prain & Burkill, Ann. Roy. Bot. Gard. (Calcutta) 14(1): 148 (1936).
*Dioscorea ochroleuca* K.Y.Guan & D.F.Chamb., Edinburgh J. Bot. 49: 85 (1992).
*Dioscorea kamoonensis* var. *brevifolia* Prain & Burkill, Ann. Roy. Bot. Gard. (Calcutta) 14(1): 148 (1936). *Dioscorea brevifolia* (Prain & Burkill) K.Y.Guan & D.F.Chamb., Edinburgh J. Bot. 49: 86 (1992).
*Dioscorea kamoonensis* var. *vera* Prain & Burkill, Ann. Roy. Bot. Gard. (Calcutta) 14(1): 149 (1936).

*Dioscorea karatana* Wilkin, Kew Bull. 55: 427 (2000).
Madagascar. 29 MDG.

*Dioscorea keduensis* Burkill ex Backer, Handb. Fl. Java 3: 114 (1924).
C. Jawa, SW. Sulawesi. 42 JAW SUL. Cl. tuber geophyte.

*Dioscorea kerrii* Prain & Burkill, J. Proc. Asiat. Soc. Bengal 10: 20 (1914).
Thailand. 41 THA. Cl. tuber geophyte.

*Dioscorea killipii* R.Knuth, Repert. Spec. Nov. Regni Veg. 22: 345 (1926).
W. South America. 83 CLM PER. Cl. tuber geophyte.

*Dioscorea kingii* R.Knuth in H.G.A.Engler (ed.), Pflanzenr., IV, 43: 289 (1924).
Pen. Malaysia. 42 MLY. Cl. tuber geophyte.
*Dioscorea harrissii* R.Knuth in H.G.A.Engler (ed.), Pflanzenr., IV, 43: 352 (1924).
*Dioscorea nurii* R.Knuth in H.G.A.Engler (ed.), Pflanzenr., IV, 43: 352 (1924).
*Dioscorea porteri* Prain & Burkill ex Ridl., Fl. Malay Penins. 4: 318 (1924).

*Dioscorea kjellbergii* R.Knuth, Repert. Spec. Nov. Regni Veg. 36: 128 (1934).
C. Sulawesi. 42 SUL. Cl. tuber geophyte.

*Dioscorea knuthiana* De Wild., Bull. Jard. Bot. État 4: 354 (1914).
Zaïre. 23 ZAI. Cl. tuber geophyte.

*Dioscorea koepperi* Standl., Publ. Field Mus. Nat. Hist., Bot. Ser. 9: 269 (1940).
Honduras. 80 HON. Cl. tuber geophyte.

*Dioscorea koyamae* Jayas., Brittonia 42: 142 (1990).
Sri Lanka. 40 SRL. Cl. tuber geophyte.

*Dioscorea kratica* Prain & Burkill, Bull. Misc. Inform. Kew 1927: 241 (1927).
Indo-China. 41 THA VIE. Cl. tuber geophyte.

*Dioscorea kuntzei* Uline ex Kuntze, Revis. Gen. Pl. 3(2): 311 (1898).
Bolivia. 83 BOL. Cl. tuber geophyte.

*Dioscorea lacerdaei* Griseb. in C.F.P.von Martius & auct. suc. (eds.), Fl. Bras. 3(1): 31 (1842). *Helmia lacerdaei* (Griseb.) Kunth, Enum. Pl. 5: 427 (1850).
N. Brazil. 84 BZN. Cl. tuber geophyte.

*Dioscorea laevis* Uline, Bot. Jahrb. Syst. 22: 425 (1896).
Nicaragua to Costa Rica. 80 COS NIC. Cl. tuber geophyte.

*Dioscorea lamprocaula* Prain & Burkill, Bull. Misc. Inform. Kew 1933: 245 (1933).
W. Malesia. 42 JAW MLY SUM. Cl. tuber geophyte.

*Dioscorea lanata* Bail, Proc. Roy. Soc. Edinburgh 12: 96 (1884).
Socotra (Haghier Hills). 24 SOC. Cl. tuber geophyte.

*Dioscorea larecajensis* Uline ex R.Knuth, Notizbl. Bot. Gart. Berlin-Dahlem 7: 195 (1917).
W. South America. 83 BOL ECU PER. Cl. tuber geophyte.
*Dioscorea helmiicarpa* R.Knuth, Repert. Spec. Nov. Regni Veg. 36: 125 (1934).

*Dioscorea lasseriana* Steyerm., Fieldiana, Bot. 28(1): 158 (1951).
Venezuela. 82 VEN. Cl. tuber geophyte.

*Dioscorea laurifolia* Wall. ex Hook.f., Fl. Brit. India 6: 293 (1892).
Pen. Thailand to Malaya, Borneo. 41 THA VIE 42 BOR MLY. Cl. tuber geophyte.
*Dioscorea laurifolia* var. *hookeri* R.Knuth in H.G.A.Engler (ed.), Pflanzenr., IV, 43: 289 (1924).

*Dioscorea lawrancei* R.Knuth, Repert. Spec. Nov. Regni Veg. 38: 118 (1935).
Colombia. 83 CLM. Cl. tuber geophyte.

*Dioscorea laxiflora* Mart. ex Griseb. in C.F.P.von Martius & auct. suc. (eds.), Fl. Bras. 3(1): 32 (1842).
S. Venezuela to Peru and Brazil. 82 VEN 83 PER 84 BZC BZE BZL BZS. Cl. tuber geophyte.
*Dioscorea laxiflora* var. *auriculata* Griseb. in C.F.P.von Martius & auct. suc. (eds.), Fl. Bras. 3(1): 33 (1842).
*Dioscorea laxiflora* var. *truncata* Griseb. in C.F.P.von Martius & auct. suc. (eds.), Fl. Bras. 3(1): 33 (1842).
*Dioscorea calystegioides* Kunth, Enum. Pl. 5: 362 (1850). *Dioscorea laxiflora* var. *calystegioides* (Kunth) Uline ex R.Knuth, Notizbl. Bot. Gart. Berlin-Dahlem 7: 214 (1917).
*Dioscorea laxiflora* var. *cincinnata* Uline ex R.Knuth, Notizbl. Bot. Gart. Berlin-Dahlem 7: 214 (1917).
*Dioscorea laxiflora* var. *cissifolia* Uline ex R.Knuth, Notizbl. Bot. Gart. Berlin-Dahlem 7: 214 (1917).
*Dioscorea laxiflora* var. *truncatolanceolata* Uline ex R.Knuth, Notizbl. Bot. Gart. Berlin-Dahlem 7: 214 (1917).

*Dioscorea lehmannii* Uline, Bot. Jahrb. Syst. 22: 430 (1896).
Colombia. 83 CLM. Cl. tuber geophyte.

*Dioscorea lepcharum* Prain & Burkill, J. Proc. Asiat. Soc. Bengal 10: 36 (1914).
E. Himalaya to Myanmar. 40 ASS BAN EHM 41 MYA. Cl. tuber geophyte.

*Dioscorea lepcharum* var. *bhamoica* Prain & Burkill, J. Proc. Asiat. Soc. Bengal 10: 36 (1914).

*Dioscorea lepida* C.V.Morton, Publ. Carnegie Inst. Wash. 461: 248 (1936).
SE. Mexico to C. America. 79 MXT 80 COS GUA HON PAN. Cl. tuber geophyte.
*Dioscorea racemosa* var. *hoffmannii* Uline, Bot. Jahrb. Syst. 22: 431 (1897).

*Dioscorea leptobotrys* Uline ex R.Knuth, Notizbl. Bot. Gart. Berlin-Dahlem 7: 218 (1917).
Brazil (São Paulo). 84 BZL. Cl. tuber geophyte.

*Dioscorea liebmannii* Uline, Bot. Jahrb. Syst. 22: 429 (1896).
S. Mexico to C. America. 79 MXS MXT 80 HON NIC PAN. Cl. tuber geophyte.

*Dioscorea lijiangensis* C.L.Long & H.Li, Ann. Bot. Fenn. 37: 193 (2000).
China (Yunnan). 36 CHC.

*Dioscorea linearicordata* Prain & Burkill, Bull. Misc. Inform. Kew 1925: 61 (1925).
SE. China. 36 CHS. Cl. tuber geophyte.

*Dioscorea lisae* Dorr & Stergios, Sida 20: 1007 (2003).
Venezuela. 82 VEN.

*Dioscorea listeri* Prain & Burkill, J. Proc. Asiat. Soc. Bengal 4: 452 (1908).
Assam. 40 ASS. Cl. tuber geophyte.

*Dioscorea litoralis* Phil., Linnaea 29: 64 (1857).
C. Chile. 85 CLC. Cl. tuber geophyte.

*Dioscorea loefgrenii* R.Knuth, Notizbl. Bot. Gart. Berlin-Dahlem 7: 187 (1917).
SE. Brazil. 84 BZL. Cl. tuber geophyte.

*Dioscorea loheri* Prain & Burkill, J. Proc. Asiat. Soc. Bengal 10: 33 (1914).
Philippines (Luzon). 42 PHI. Cl. tuber geophyte.
*Dioscorea wilkesii* Uline ex R.Knuth in H.G.A.Engler (ed.), Pflanzenr., IV, 43: 271 (1924).

*Dioscorea longicuspis* R.Knuth, Notizbl. Bot. Gart. Berlin-Dahlem 11: 1059 (1934).
Tanzania. 25 TAN. Cl. tuber geophyte.
*Dioscorea schimperiana* var. *nigrescens* Uline ex R.Knuth in H.G.A.Engler (ed.), Pflanzenr., IV, 43: 256 (1924).

*Dioscorea longipes* Phil., Anales Univ. Chile 93: 6 (1896).
Chile (Maule). 85 CLC. Tuber geophyte.

*Dioscorea longirhiza* Caddick & Wilkin, Taxon 51: 112 (2002).
SW. Mexico. 79 MXS. Cl. tuber geophyte.
*Nanarepenta guerrerensis* Matuda, Cact. Suc. Mex. 19: 70 (1974).

*Dioscorea longituba* Uline in H.G.A.Engler & K.A.E.Prantl, Nat. Pflanzenfam., Nachtr. 1: 86 (1897).
Mexico (México State, Hidalgo, Nayarit). 79 MXC MXE MXS. Cl. tuber geophyte.

*Dioscorea lundii* Uline ex R.Knuth, Notizbl. Bot. Gart. Berlin-Dahlem 7: 198 (1917).
SE. Brazil. 84 BZL. Cl. tuber geophyte.

*Dioscorea luzonensis* Schauer, Nov. Actorum Acad. Caes. Leop.-Carol. Nat. Cur. 19(Suppl. 1): 444 (1843).
Philippines. 42 PHI. Cl. tuber geophyte.

*Dioscorea macbrideana* R.Knuth, Repert. Spec. Nov. Regni Veg. 28: 87 (1930).
Peru. 83 PER. Cl. tuber geophyte.
*Dioscorea ramonensis* R.Knuth, Repert. Spec. Nov. Regni Veg. 39: 95 (1931).

*Dioscorea macrantha* Uline ex R.Knuth, Notizbl. Bot. Gart. Berlin-Dahlem 7: 198 (1917).
Bolivia to S. Brazil. 83 BOL 84 BZS. Cl. tuber geophyte.

*Dioscorea macrothyrsa* Uline in H.G.A.Engler & K.A.E.Prantl, Nat. Pflanzenfam., Nachtr. 1: 87 (1897).
NE. Brazil. 82 GUY? 84 BZE. Cl. tuber geophyte.

*Dioscorea macvaughii* B.G.Schub., in Fl. Novo-Galiciana 15: 369 (1989).
SW. Mexico. 79 MXS. Cl. tuber geophyte.

*Dioscorea madecassa* H.Perrier, Mém. Soc. Linn. Normandie, Bot. 1(1): 24 (1928).
Madagascar. 29 MDG. Cl. tuber geophyte.

*Dioscorea madiunensis* Prain & Burkill, Bull. Misc. Inform. Kew 1925: 63 (1925).
Jawa. 42 JAW. Cl. tuber geophyte.
*Dioscorea gedensis* Prain & Burkill, Bull. Misc. Inform. Kew 1925: 64 (1925).

*Dioscorea maianthemoides* Uline ex R.Knuth, Notizbl. Bot. Gart. Berlin-Dahlem 7: 188 (1917).
Brazil (Minas Gerais to Goiás). 84 BZC BZL. Tuber geophyte.

*Dioscorea mamillata* Jum. & H.Perrier, Ann. Inst. Bot.-Géol. Colon. Marseille, II, 8: 422 (1910).
NW. Madagascar. 29 MDG. Tuber geophyte.

*Dioscorea mandonii* Rusby, Bull. Torrey Bot. Club 29: 701 (1902).
Bolivia. 83 BOL. Cl. tuber geophyte.
*Dioscorea arcuata* Rusby, Bull. New York Bot. Gard. 4: 460 (1907).
*Dioscorea ancistrocarpa* Uline ex R.Knuth in H.G.A. Engler (ed.), Pflanzenr., IV, 43: 105 (1924), nom. inval.

*Dioscorea mangenotiana* J.Miège, Bull. Inst. Fondam. Afrique Noire, Sér. A., Sci. Nat. 20: 40 (1958).
W. & WC. Trop. Africa. 22 BEN IVO LBR NGA TOG 23 CAF ZAI. Cl. tuber geophyte.

*Dioscorea mantigueirensis* R.Knuth, Notizbl. Bot. Gart. Berlin-Dahlem 7: 192 (1917).
SE. Brazil. 84 BZL. Cl. tuber geophyte.

*Dioscorea margarethia* G.M.Barroso, E.F.Guim. & Sucre, Loefgrenia 49: 2 (1970).
Brazil (Rio de Janeiro). 84 BZL. Cl. tuber geophyte.

*Dioscorea marginata* Griseb. in C.F.P.von Martius & auct. suc. (eds.), Fl. Bras. 3(1): 37 (1842).
Brazil. 84 BZC BZE BZL BZN BZS. Cl. tuber geophyte.
*Dioscorea cynanchifolia* Griseb., Vidensk. Meddel. Naturhist. Foren. Kjøbenhavn 1875: 156 (1875).
*Dioscorea albinervia* R.Knuth in H.G.A.Engler (ed.), Pflanzenr., IV, 43: 74 (1924).

*Dioscorea martensis* R.Knuth, Notizbl. Bot. Gart. Berlin-Dahlem 7: 214 (1917).
Colombia. 83 CLM. Cl. tuber geophyte.

*Dioscorea martiana* Griseb. in C.F.P.von Martius & auct. suc. (eds.), Fl. Bras. 3(1): 44 (1842). *Dioscorea martiana* var. *genuina* R.Knuth in H.G.A.Engler (ed.), Pflanzenr., IV, 43: 218 (1924), nom. inval.

Brazil. 84 BZC BZE BZL BZS. Cl. tuber geophyte.
*Dioscorea leptostachya* Gardner, London J. Bot. 1: 534 (1842). *Dioscorea martiana* var. *leptostachya* (Gardner) Uline ex R.Knuth, Notizbl. Bot. Gart. Berlin-Dahlem 7: 208 (1917).
*Dioscorea micrantha* Kunth, Enum. Pl. 5: 327 (1850).
*Dioscorea fluminensis* R.Knuth, Notizbl. Bot. Gart. Berlin-Dahlem 7: 210 (1917).
*Dioscorea martiana* var. *caudata* R.Knuth in H.G.A. Engler (ed.), Pflanzenr., IV, 43: 218 (1924).
*Dioscorea martiana* var. *pedicellata* R.Knuth in H.G.A.Engler (ed.), Pflanzenr., IV, 43: 218 (1924).

*Dioscorea martini* Prain & Burkill, J. Proc. Asiat. Soc. Bengal 10: 18 (1914).
SC. China. 36 CHC. Cl. tuber geophyte.

*Dioscorea matagalpensis* Uline, Bot. Jahrb. Syst. 22: 432 (1896).
SE. Mexico to Colombia. 79 MXT 80 BLZ COS GUA HON NIC PAN 83 CLM. Cl. tuber geophyte.
*Dioscorea yucatanensis* Uline, Publ. Field Columbian Mus., Bot. Ser. 1: 16 (1895).

*Dioscorea matudae* O.Téllez & B.G.Schub., Ann. Missouri Bot. Gard. 74: 539 (1987).
NE. Mexico. 79 MXE. Cl. tuber geophyte.

*Dioscorea megacarpa* Gleason, Bull. Torrey Bot. Club 52: 184 (1925).
S. Trop. America. 82 FRG GUY SUR 83 PER. Cl. tuber geophyte.
*Dioscorea truncata* R.H.Schomb. ex Prain, Bull. Misc. Inform. Kew 1916: 194 (1916), nom. illeg.

*Dioscorea megalantha* Griseb., Abh. Königl. Ges. Wiss. Göttingen 24: 323 (1879).
NW. Argentina. 85 AGW. Cl. tuber geophyte.
*Dioscorea megalantha* var. *lilloi* Hauman, Anales Mus. Nac. Hist. Nat. Buenos Aires 27: 474 (1915).
*Dioscorea lilloi* (Hauman) Hauman, Anales Mus. Nac. Hist. Nat. Buenos Aires 29: 431 (1917).

*Dioscorea melanophyma* Prain & Burkill, J. Proc. Asiat. Soc. Bengal 4: 452 (1908).
Himalaya to SC. China. 36 CHC CHT 40 ASS BAN EHM NEP WHM. Cl. tuber geophyte.
*Dioscorea tenii* R.Knuth in H.G.A.Engler (ed.), Pflanzenr., IV, 43: 142 (1924).

*Dioscorea melastomatifolia* Uline ex Prain, Bull. Misc. Inform. Kew 1916: 194 (1916).
French Guiana to N. Brazil. 82 FRG 84 BZN. Cl. tuber geophyte.

*Dioscorea membranacea* Pierre ex Prain & Burkill, J. Proc. Asiat. Soc. Bengal 10: 13 (1914).
Indo-China to Pen. Malaysia. 41 CBD MYA THA VIE 42 MLY. Rhizome geophyte.

*Dioscorea menglaensis* H.Li, in Fl. Yunnanica 3: 744 (1983).
China (W. Yunnan). 36 CHC. Cl. tuber geophyte.

*Dioscorea meridensis* Kunth, Enum. Pl. 5: 334 (1850).
W. South America to Venezuela. 82 VEN 83 CLM PER. Cl. tuber geophyte.

*Dioscorea merrillii* Prain & Burkill, Leafl. Philipp. Bot. 5: 1598 (1913).
Philippines. 42 PHI. Cl. tuber geophyte.

*Dioscorea mesoamericana* O.Téllez & Mart.-Rodr., Novon 3: 204 (1993).
S. Mexico. 79 MXS MXT. Cl. tuber geophyte.

*Dioscorea mexicana* Scheidw., Hort. Belge 4: 99 (1837).
Mexico to C. America. 79 MXE MXG MXS MXT 80
BLZ COS GUA HON NIC PAN. Caudex geophyte.
*Dioscorea macrostachya* Benth., Pl. Hartw.: 73
(1841). *Testudinaria macrostachya* (Benth.)
G.D.Rowley, Natl. Cact. Succ. J. 28: 6 (1973).
*Dioscorea macrophylla* M.Martens & Galeotti, Bull.
Acad. Roy. Sci. Bruxelles 9(2): 392 (1842).
*Dioscorea deppei* Schiede ex Schltdl., Bot. Zeitung
(Berlin) 1: 890 (1843).
*Dioscorea bilbergiana* Kunth, Enum. Pl. 5: 354 (1850).
*Dioscorea leiboldiana* Kunth, Enum. Pl. 5: 355 (1850).
*Dioscorea propinqua* Hemsl., Biol. Cent.-Amer., Bot.
3: 359 (1884).
*Testudinaria cocolmeca* Procop., Bot. Centralbl. 49:
201 (1892).
*Dioscorea astrostigma* Uline, Bot. Jahrb. Syst. 22: 431
(1896).
*Dioscorea macrostachya* var. *sessiliflora* Uline, Bot.
Jahrb. Syst. 22: 424 (1897). *Dioscorea mexicana*
var. *sessiliflora* (Uline) Matuda, Anales Inst. Biol.
Univ. Nac. México 24: 285 (1954).
*Dioscorea tuerckheimii* R.Knuth, Notizbl. Bot. Gart.
Berlin-Dahlem 7: 203 (1917).
*Dioscorea anconensis* R.Knuth, Repert. Spec. Nov.
Regni Veg. 28: 82 (1930).
*Dioscorea deamii* Matuda, Anales Inst. Biol. Univ.
Nac. México 24: 60 (1953).

*Dioscorea microbotrya* Griseb., Abh. Königl. Ges. Wiss.
Göttingen 24: 322 (1879).
Brazil to N. Argentina. 84 BZC BZE BZL BZS 85
AGE AGW PAR URU. Cl. tuber geophyte.
*Dioscorea microbotrya* var. *grandifolia* Hauman,
Anales Mus. Nac. Hist. Nat. Buenos Aires 27: 464
(1915).
*Dioscorea filirachis* R.Knuth in H.G.A.Engler (ed.),
Pflanzenr., IV, 43: 128 (1924), pro syn.
*Dioscorea gibertii* R.Knuth in H.G.A.Engler (ed.),
Pflanzenr., IV, 43: 58 (1924).
*Dioscorea tweediei* R.Knuth in H.G.A.Engler (ed.),
Pflanzenr., IV, 43: 58 (1924).

*Dioscorea microcephala* Uline in H.G.A.Engler &
K.A.E.Prantl, Nat. Pflanzenfam., Nachtr. 1: 85 (1897).
S. Brazil. 84 BZS. Cl. tuber geophyte.

*Dioscorea microura* R.Knuth, Notizbl. Bot. Gart.
Berlin-Dahlem 7: 196 (1917).
Guyana. 82 GUY. Cl. tuber geophyte.

*Dioscorea militaris* C.B.Rob., Proc. Amer. Acad. Arts
29: 324 (1894).
NE. & SW. Mexico. 79 MXE MXS. Cl. tuber geophyte.

*Dioscorea mindanaensis* R.Knuth in H.G.A.Engler
(ed.), Pflanzenr., IV, 43: 271 (1924).
Philippines (Mindanao). 42 PHI. Cl. tuber geophyte.

*Dioscorea minima* C.B.Rob. & Seaton, Proc. Amer.
Acad. Arts 28: 115 (1893).
SW. Mexico. 79 MXS. Tuber geophyte.
*Dioscorea pumila* Sessé & Moç., Fl. Mexic., ed. 2: 231
(1894).

*Dioscorea minutiflora* Engl., Bot. Jahrb. Syst. 7: 332
(1886). *Dioscorea praehensilis* var. *minutiflora*
(Engl.) Baker in D.Oliver & auct. suc. (eds.), Fl. Trop.
Afr. 7: 417 (1898).
W. Trop. Africa to Uganda, Madagascar. 22 BEN GUI
IVO LBR NGA SEN SIE TOG 23 CAB CAF CMN
EQG GAB ZAI 25 UGA 29 MDG. Cl. tuber
geophyte.

*Dioscorea multiflora* Engl. ex Pax, Bot. Jahrb. Syst.
15: 146 (1892), orth. var.
*Dioscorea acarophyta* De Wild., Compt. Rend. Hebd.
Séances Acad. Sci. 139: 552 (1904).
*Dioscorea brevispicata* De Wild., Ann. Mus. Congo
Belge, Bot., V, 3: 358 (1912).
*Dioscorea ealensis* De Wild., Ann. Mus. Congo Belge,
Bot., V, 3: 359 (1912).
*Dioscorea lilela* De Wild., Ann. Mus. Congo Belge,
Bot., V, 3: 365 (1912).
*Dioscorea litoie* De Wild., Ann. Mus. Congo Belge,
Bot., V, 3: 364 (1912).
*Dioscorea armata* De Wild., Bull. Jard. Bot. État 4:
339 (1914).
*Dioscorea ekolo* De Wild., Bull. Jard. Bot. État 4: 341
(1914).
*Dioscorea engbo* De Wild., Bull. Jard. Bot. État 4: 342
(1914).
*Dioscorea pynaertioides* De Wild., Bull. Jard. Bot. État
4: 314 (1914).
*Dioscorea grandibulbosa* R.Knuth in H.G.A.Engler
(ed.), Pflanzenr., IV, 43: 304 (1924).
*Dioscorea hystrix* R.Knuth in H.G.A.Engler (ed.),
Pflanzenr., IV, 43: 304 (1924).
*Dioscorea pendula* R.Knuth in H.G.A.Engler (ed.),
Pflanzenr., IV, 43: 305 (1924), nom. illeg.

**Dioscorea mitis** C.V.Morton, Publ. Carnegie Inst. Wash.
461: 247 (1936).
SW. Mexico. 79 MXS. Cl. tuber geophyte.

**Dioscorea mitoensis** R.Knuth, Repert. Spec. Nov. Regni
Veg. 28: 84 (1930).
Peru. 83 PER. Cl. tuber geophyte.

**Dioscorea modesta** Phil., Linnaea 33: 256 (1865).
C. Chile. 85 CLC. Cl. tuber geophyte.
*Dioscorea parvifolia* Phil., Anales Univ. Chile 93: 8
(1896). *Dioscorea aristolochiifolia* var. *parvifolia*
(Phil.) L.E.Navas, Anales Acad. Chilena Ci. Nat.
1968: 55 (1968).

**Dioscorea mollis** Kunth, Enum. Pl. 5: 369 (1850).
SE. & S. Brazil. 84 BZL BZS. Cl. tuber geophyte.
*Dioscorea pachycarpa* Kunth, Enum. Pl. 5: 370
(1850). *Dioscorea mollis* var. *pachycarpa* (Kunth)
Uline ex R.Knuth, Notizbl. Bot. Gart. Berlin-
Dahlem 7: 201 (1917).

**Dioscorea monadelpha** (Kunth) Griseb., Vidensk.
Meddel. Naturhist. Foren. Kjøbenhavn 1875: 164
(1875).
Peru, SE. & S. Brazil. 83 PER 84 BZL BZS. Cl. tuber
geophyte.
*Helmia monodelpha* Kunth, Enum. Pl. 5: 421 (1850).
*Dioscorea monadelphoides* J.F.Macbr., Publ. Field
Mus. Nat. Hist., Bot. Ser. 11: 12 (1931), nom. illeg.
*Dioscorea friesii* R.Knuth, Repert. Spec. Nov. Regni
Veg. 21: 77 (1925).
*Dioscorea similis* R.Knuth, Repert. Spec. Nov. Regni
Veg. 21: 78 (1925).
*Dioscorea longirachis* R.Knuth, Repert. Spec. Nov.
Regni Veg. 30: 159 (1932). *Dioscorea*
*monadelphoides* var. *longirachis* (R.Knuth) Ayala,
Diosc. Peru: 38 (1998).

**Dioscorea ×monandra** Hauman, Anales Mus. Nac. Hist.
Nat. Buenos Aires 27: 485 (1916). *D. cieneyensis* ×
*D. glomerulata*.
NW. Argentina. 85 AGW. Cl. tuber geophyte.

**Dioscorea morelosana** (Uline) Matuda, Anales Inst.
Biol. Univ. Nac. México 24: 61 (1953).

C. & SW. Mexico. 79 MXC MXS. Cl. tuber geophyte.
*Dioscorea lobata* var. *morelosana* Uline, Proc. Amer. Acad. Arts 35: 323 (1900).

**Dioscorea moritziana** (Kunth) R.Knuth, Notizbl. Bot. Gart. Berlin-Dahlem 7: 197 (1917).
Venezuela. 82 VEN. Cl. tuber geophyte.
*Helmia moritziana* Kunth, Enum. Pl. 5: 422 (1850).

**Dioscorea mosqueirensis** R.Knuth, Repert. Spec. Nov. Regni Veg. 29: 92 (1931).
Brazil (Pará). 84 BZN. Cl. tuber geophyte.

**Dioscorea moultonii** Prain & Burkill, Bull. Misc. Inform. Kew 1925: 62 (1925).
Borneo (Sarawak). 42 BOR. Cl. tuber geophyte.

**Dioscorea moyobambensis** R.Knuth, Notizbl. Bot. Gart. Berlin-Dahlem 7: 185 (1917).
Peru. 83 PER. Cl. tuber geophyte.
*Dioscorea tambillensis* R.Knuth, Repert. Spec. Nov. Regni Veg. 28: 81 (1930).

**Dioscorea mucronata** Uline ex R.Knuth, Notizbl. Bot. Gart. Berlin-Dahlem 7: 188 (1917).
NE. Brazil. 84 BZE. Cl. tuber geophyte.

**Dioscorea multiflora** Mart. ex Griseb. in C.F.P.von Martius & auct. suc. (eds.), Fl. Bras. 3(1): 35 (1842).
*Helmia multiflora* (Mart. ex Griseb.) Kunth, Enum. Pl. 5: 481 (1850).
Brazil to Paraguay. 84 BZC BZL BZS 85 AGE PAR. Cl. tuber geophyte.
*Dioscorea multiflora* var. *parvifolia* Griseb. in C.F.P.von Martius & auct. suc. (eds.), Fl. Bras. 3(1): 35 (1842).
*Dioscorea macrocapsa* Uline in H.G.A.Engler & K.A.E. Prantl, Nat. Pflanzenfam., Nachtr. 1: 83 (1897).
*Dioscorea concepcionis* Chodat & Hassl., Bull. Herb. Boissier, II, 3: 1111 (1903).
*Dioscorea multiflora* var. *asuncianensis* Uline ex R.Knuth, Notizbl. Bot. Gart. Berlin-Dahlem 7: 190 (1917).
*Dioscorea multiflora* var. *asuncionensis* Uline, Notizbl. Bot. Gart. Berlin-Dahlem 7: 190 (1917).
*Dioscorea multiflora* var. *loefgrenii* Uline ex R.Knuth, Notizbl. Bot. Gart. Berlin-Dahlem 7: 190 (1917).
*Dioscorea multiflora* var. *lofgrenii* R.Knuth, Notizbl. Bot. Gart. Berlin-Dahlem 7: 190 (1917).
*Dioscorea niederleinii* R.Knuth, Notizbl. Bot. Gart. Berlin-Dahlem 7: 193 (1917).
*Dioscorea multiflora* var. *concepcionis* Pellegr., Bull. Soc. Bot. Genève, II, 10: 384 (1919).

**Dioscorea multiloba** Kunth, Enum. Pl. 5: 376 (1850).
S. Africa. 27 CPP NAT SWZ. Cl. tuber geophyte.
*Dioscorea digitaria* R.Knuth in H.G.A.Engler (ed.), Pflanzenr., IV, 43: 184 (1924).

**Dioscorea multinervis** Benth., Pl. Hartw.: 52 (1840).
SW. & C. Mexico. 79 MXC MXS. Tuber geophyte.
*Dioscorea nana* Schltdl., Linnaea 18: 112 (1844), nom. illeg. *Dioscorea schlechtendalii* Kunth, Enum. Pl. 5: 411 (1850).
*Nanarepenta tolucana* Matuda, Anales Inst. Biol. Univ. Nac. México 32: 144 (1962). *Dioscorea tolucana* (Matuda) Caddick & Wilkin, Taxon 51: 112 (2002).

**Dioscorea mundtii** Baker, J. Bot. 27: 1 (1889).
Cape Prov. 27 CPP. Cl. tuber geophyte.

**Dioscorea nako** H.Perrier, Mém. Soc. Linn. Normandie, Bot. 1(1): 30 (1928).
SW. Madagascar. 29 MDG. Cl. tuber geophyte.

**Dioscorea namorokensis** Wilkin, Kew Bull. 57: 902 (2002).
Madagascar. 29 MDG.

**Dioscorea nana** Poepp., Fragm. Syn. Pl.: 12 (1833).
SC. Chile. 85 CLC. Tuber geophyte.

**Dioscorea nanlaensis** H.Li, in Fl. Yunnanica 3: 739 (1983).
China (Yunnan). 36 CHC. Cl. tuber geophyte.

**Dioscorea natalensis** R.Knuth in H.G.A.Engler (ed.), Pflanzenr., IV, 43: 94 (1924).
KwaZulu-Natal. 27 NAT. Cl. tuber geophyte.

**Dioscorea natalia** Hammel, Novon 10: 378 (2000).
Costa Rica. 80 COS. Cl.

**Dioscorea neblinensis** Maguire & Steyerm., Mem. New York Bot. Gard. 51: 108 (1989).
Venezuela (Amazonas) to Brazil (Serra Pirapucú). 82 VEN 84 BZN. Cl. tuber geophyte.

**Dioscorea nelsonii** Uline ex R.Knuth, Notizbl. Bot. Gart. Berlin-Dahlem 7: 202 (1917).
S. Mexico to Guatemala. 79 MXS MXT 80 GUA. Cl. tuber geophyte.

**Dioscorea nematodes** Uline ex R.Knuth, Notizbl. Bot. Gart. Berlin-Dahlem 7: 193 (1917).
Mexico (Guanajuato). 79 MXE. Cl. tuber geophyte.

**Dioscorea nervata** R.Knuth, Repert. Spec. Nov. Regni Veg. 22: 345 (1926).
W. South America. 83 CLM PER. Cl. tuber geophyte.

**Dioscorea nervosa** Phil., Anales Univ. Chile 93: 20 (1902).
*Dioscorea reticulata* var. *nervosa* (Phil.) L.E.Navas, Anales Acad. Chilena Ci. Nat. 1968: 54 (1968).
S. Chile. 85 CLS. Cl. tuber geophyte.

**Dioscorea nicolasensis** R.Knuth, Repert. Spec. Nov. Regni Veg. 29: 94 (1931).
Peru. 83 PER. Cl. tuber geophyte.

**Dioscorea nieuwenhuisii** Prain & Burkill, Bull. Misc. Inform. Kew 1925: 65 (1925).
EC. Borneo. 42 BOR. Cl. tuber geophyte.

**Dioscorea nipensis** R.A.Howard, J. Arnold Arbor. 28: 119 (1947).
Cuba. 81 CUB. Cl. tuber geophyte.
*Dioscorea linearis* Griseb., Cat. Pl. Cub.: 251 (1866), nom. illeg. *Dioscorea grisebachii* Britton ex Léon, Contr. Ocas. Mus. Hist. Nat. Colegio "De La Salle" 8: 321 (1946), nom. illeg. *Rajania linearis* R.A.Howard, J. Arnold Arbor. 28: 118 (1947).
*Dioscorea ravenii* Ayala, Phytologia 55: 296 (1984).
*Dioscorea montecristina* Hadac, Folia Geobot. Phytotax. 5: 430 (1970).

**Dioscorea nipponica** Makino, Ill. Fl. Jap. 1(7): 2 (1891).
C. China to N. & C. Japan. 31 AMU KHA PRM 36 CHC CHI CHM CHN CHQ CHS 38 JAP KOR. Cl. rhizome geophyte.
*Dioscorea acerifolia* var. *rosthornii* Diels, Bot. Jahrb. Syst. 29: 261 (1900). *Dioscorea nipponica* var. *rosthornii* (Diels) Prain & Burkill, J. Proc. Asiat. Soc. Bengal 10: 14 (1914). *Dioscorea nipponica* subsp. *rosthornii* (Diels) C.T.Ting, Acta Phytotax. Sin. 17(3): 70 (1979).
*Dioscorea acerifolia* Uline ex Prain & Burkill, J. Asiat. Soc. Bengal, Pt. 2, Nat. Hist. 73(2 Suppl.): 7 (1904 publ. 1905).
*Dioscorea nipponica* var. *jamesii* Prain & Burkill, J. Proc. Asiat. Soc. Bengal 10: 14 (1914).

*Dioscorea giraldii* R.Knuth in H.G.A.Engler (ed.), Pflanzenr., IV, 43: 315 (1924).

**Dioscorea nitens** Prain & Burkill, J. Proc. Asiat. Soc. Bengal 10: 18 (1914).
China (Yunnan). 36 CHC. Cl. tuber geophyte.

**Dioscorea nuda** R.Knuth in H.G.A.Engler (ed.), Pflanzenr., IV, 43: 351 (1924).
SE. Brazil. 84 BZL. Cl. tuber geophyte.

**Dioscorea nummularia** Lam., Encycl. 3: 231 (1789).
Andaman Is. to Papuasia. 41 AND VIE 42 BOR JAW MOL PHI SUL 43 BIS NWG 50 QLD? (60) fij (61) sci (62) crl. Cl. tuber geophyte.
*Dioscorea pirita* Nadeaud, Énum. Pl. Tahiti: 35 (1873).
*Dioscorea seemannii* Prain & Burkill, J. Proc. Asiat. Soc. Bengal 10: 34 (1914).
*Dioscorea angulata* R.Knuth in H.G.A.Engler (ed.), Pflanzenr., IV, 43: 283 (1924).
*Dioscorea glaucoidea* R.Knuth in H.G.A.Engler (ed.), Pflanzenr., IV, 43: 284 (1924).
*Dioscorea lufensis* R.Knuth in H.G.A.Engler (ed.), Pflanzenr., IV, 43: 272 (1924).
*Dioscorea nummularia* var. *lata* R.Knuth in H.G.A.Engler (ed.), Pflanzenr., IV, 43: 283 (1924).
*Dioscorea palauensis* R.Knuth in H.G.A.Engler (ed.), Pflanzenr., IV, 43: 191 (1924).
*Dioscorea raymundii* R.Knuth in H.G.A.Engler (ed.), Pflanzenr., IV, 43: 191 (1924).
*Dioscorea palopoensis* R.Knuth, Repert. Spec. Nov. Regni Veg. 36: 128 (1934).

**Dioscorea nutans** R.Knuth in H.G.A.Engler (ed.), Pflanzenr., IV, 43: 182 (1924).
Brazil (Minas Gerais). 84 BZL. Cl. tuber geophyte.

**Dioscorea oaxacensis** Uline in H.G.A.Engler & K.A.E.Prantl, Nat. Pflanzenfam., Nachtr. 1: 86 (1897).
SW. Mexico. 79 MXS. Cl. tuber geophyte.

**Dioscorea obcuneata** Hook.f., Fl. Brit. India 6: 293 (1892).
Sri Lanka. 40 SRL. Cl. tuber geophyte.

**Dioscorea oblonga** Gleason, Bull. Torrey Bot. Club 52: 182 (1925).
Guyana to Suriname. 82 GUY SUR. Cl. tuber geophyte.

**Dioscorea oblongifolia** Rusby, Bull. New York Bot. Gard. 6: 492 (1910).
Bolivia. 83 BOL. Cl. tuber geophyte.

**Dioscorea obtusifolia** Hook. & Arn., Bot. Beechey Voy.: 48 (1830).
C. Chile. 85 CLC. Cl. tuber geophyte.
*Dioscorea obtusifolia* var. *philippii* Uline ex R.Knuth, Notizbl. Bot. Gart. Berlin-Dahlem 7: 205 (1917).

**Dioscorea olfersiana** Klotzsch ex Griseb. in C.F.P.von Martius & auct. suc. (eds.), Fl. Bras. 3(1): 38 (1842).
Brazil to Peru. 83 PER 84 BZC BZL BZS. Cl. tuber geophyte.

**Dioscorea oligophylla** Phil., Linnaea 33: 255 (1865).
S. Chile. 85 CLS. Tuber geophyte.

**Dioscorea omiltemensis** O.Téllez, Novon 12: 441 (2002).
Mexico (Guerrero). 79 MXS. Cl. tuber geophyte.

**Dioscorea opaca** R.Knuth in H.G.A.Engler (ed.), Pflanzenr., IV, 43: 283 (1924).
New Guinea. 43 NWG. Cl. tuber geophyte.
*Dioscorea carrii* R.Knuth, Repert. Spec. Nov. Regni Veg. 42. 163 (1937).

*Dioscorea morobeensis* R.Knuth, Repert. Spec. Nov. Regni Veg. 42: 164 (1937).

**Dioscorea oppositiflora** Griseb. in C.F.P.von Martius & auct. suc. (eds.), Fl. Bras. 3(1): 46 (1842).
Brazil (Rio de Janeiro). 84 BZL. Cl. tuber geophyte.

**Dioscorea oppositifolia** L., Sp. Pl.: 1033 (1753).
S. India, Sri Lanka, SE. Bangladesh. 40 BAN IND SRL. Cl. tuber geophyte.
*Dioscorea opposita* Thunb., Fl. Jap.: 151 (1784).
*Dioscorea oppositifolia* var. *dukhunensis* Prain & Burkill, J. Proc. Asiat. Soc. Bengal 10: 30 (1914).
*Dioscorea oppositifolia* var. *linnaei* Prain & Burkill, J. Proc. Asiat. Soc. Bengal 10: 30 (1914).
*Dioscorea oppositifolia* var. *thwaitesii* Prain & Burkill, J. Proc. Asiat. Soc. Bengal 10: 30 (1914).

**Dioscorea orbiculata** Hook.f., Fl. Brit. India 6: 292 (1892).
Pen. Malaysia to Sumatera. 42 MLY SUM. Cl. tuber geophyte.

**Dioscorea oreodoxa** B.G.Schub., in Fl. Novo-Galiciana 15: 378 (1989).
SW. Mexico (Colima). 79 MXS. Cl. tuber geophyte.

**Dioscorea organensis** R.Knuth in H.G.A.Engler (ed.), Pflanzenr., IV, 43: 106 (1924).
SE. Brazil. 84 BZL. Cl. tuber geophyte.

**Dioscorea orientalis** (J.Thiébaut) Caddick & Wilkin, Taxon 51: 112 (2002).
Lebanon to Israel. 34 LBS PAL. Cl. tuber geophyte.
*Tamus orientalis* J.Thiébaut, Bull. Soc. Bot. France 81: 119 (1934).

**Dioscorea orizabensis** Uline in H.G.A.Engler & K.A.E.Prantl, Nat. Pflanzenfam., Nachtr. 1: 86 (1897).
Mexico (Jalisco, Veracruz). 79 MXG MXS. Cl. tuber geophyte.

**Dioscorea orthogoneura** Uline ex Hochr., Bull. New York Bot. Gard. 6: 267 (1910).
Venezuela to Bolivia. 82 GUY VEN 83 BOL 84 BZC BZE BZL. Cl. tuber geophyte.
*Dioscorea orthogoneura* var. *acutissima* Uline ex R.Knuth, Notizbl. Bot. Gart. Berlin-Dahlem 7: 216 (1917).
*Dioscorea orthogoneura* var. *brevispicata* Uline ex R.Knuth, Notizbl. Bot. Gart. Berlin-Dahlem 7: 216 (1917).
*Dioscorea orthogoneura* var. *meiapontensis* Uline ex R.Knuth, Notizbl. Bot. Gart. Berlin-Dahlem 7: 216 (1917).

**Dioscorea oryzetorum** Prain & Burkill, Bull. Misc. Inform. Kew 1927: 242 (1927).
Indo-China to Pen. Malaysia. 41 THA VIE 42 MLY. Cl. tuber geophyte.

**Dioscorea ovalifolia** R.Knuth, Notizbl. Bot. Gart. Berlin-Dahlem 7: 200 (1917).
Bolivia. 83 BOL. Cl. tuber geophyte.

**Dioscorea ovata** Vell., Fl. Flumin. 10: t. 117 (1831).
Brazil to Bolivia. 83 BOL PER? 84 BZC BZE BZL BZN BZS 85 AGE? PAR. Cl. tuber geophyte.
*Dioscorea adenocarpa* Mart. ex Griseb. in C.F.P.von Martius & auct. suc. (eds.), Fl. Bras. 3(1): 29 (1842). *Helmia adenocarpa* (Mart. ex Griseb.) Kunth, Enum. Pl. 5: 425 (1850).
*Dioscorea adenocarpa* var. *balansae* Uline ex R.Knuth in H.G.A.Engler (ed.), Pflanzenr., IV, 43: 54 (1924).

*Dioscorea adenocarpa* var. *chartacea* Uline ex
R.Knuth in H.G.A.Engler (ed.), Pflanzenr., IV, 43:
54 (1924).

**Dioscorea ovinala** Baker, J. Bot. 20: 269 (1882).
W. Madagascar. 29 MDG. Cl. tuber geophyte.
*Dioscorea velutina* Jum. & H.Perrier, Ann. Inst. Bot.-
Géol. Colon. Marseille, II, 8: 420 (1910).

**Dioscorea palawana** Prain & Burkill, Bull. Misc.
Inform. Kew 1925: 59 (1925).
Philippines (Palawan). 42 PHI. Cl. rhizome geophyte.

**Dioscorea paleata** Burkill, Bull. Jard. Bot. État 15: 360
(1939).
Zaïre. 23 ZAI. Cl. tuber geophyte.

**Dioscorea pallens** Schltdl., Linnaea 17: 610 (1843).
Mexico (Veracruz). 79 MXG. Cl. tuber geophyte.
*Dioscorea polygonoides* M.Martens & Galeotti, Bull.
Acad. Roy. Sci. Bruxelles 9(2): 393 (1842), nom.
illeg.

**Dioscorea pallidinervia** R.Knuth in H.G.A.Engler (ed.),
Pflanzenr., IV, 43: 75 (1924).
S. Brazil. 84 BZS. Cl. tuber geophyte.

**Dioscorea palmeri** R.Knuth, Notizbl. Bot. Gart. Berlin-
Dahlem 7: 203 (1917). *Dioscorea macrostachya* var.
*palmeri* (R.Knuth) C.V.Morton, Publ. Carnegie Inst.
Wash. 461: 249 (1936).
SW. Mexico. 79 MXS. Caudex geophyte.

**Dioscorea panamensis** R.Knuth in H.G.A.Engler (ed.),
Pflanzenr., IV, 43: 109 (1924).
Costa Rica to Panama, S. Venezuela. 80 COS PAN 82
VEN. Cl. tuber geophyte.

**Dioscorea panthaica** Prain & Burkill, J. Asiat. Soc.
Bengal, Pt. 2, Nat. Hist. 73(Suppl.): 6 (1904).
SC. China (to NW. Hunan), Thailand. 36 CHC CHS
41 THA. Cl. rhizome geophyte.
*Dioscorea biserialis* Prain & Burkill, Bull. Misc.
Inform. Kew 1925: 58 (1925).

**Dioscorea pantojensis** R.Knuth, Repert. Spec. Nov.
Regni Veg. 40: 223 (1936).
C. Mexico. 79 MXC. Cl. tuber geophyte.

**Dioscorea paradoxa** Prain & Burkill, Bull. Misc.
Inform. Kew 1927: 246 (1927).
Indo-China. 41 THA VIE. Cl. tuber geophyte.

**Dioscorea pavonii** Uline ex R.Knuth, Notizbl. Bot. Gart.
Berlin-Dahlem 7: 215 (1917).
Peru. 83 PER. Cl. tuber geophyte.

**Dioscorea pedicellata** Phil., Anales Univ. Chile 1873:
540 (1873).
C. Chile. 85 CLC. Cl. tuber geophyte.

**Dioscorea pencana** Phil., Anales Univ. Chile 93: 7 (1896).
SC. Chile. 85 CLC. Cl. tuber geophyte.

**Dioscorea pendula** Poepp. ex Kunth, Enum. Pl. 5: 367
(1850). *Dioscorea amaranthoides* var. *pendula*
(Poepp. ex Kunth) Uline ex R.Knuth, Notizbl. Bot.
Gart. Berlin-Dahlem 7: 215 (1917).
N. Brazil to Peru. (2) 83 PER 84 BZN. Cl. tuber
geophyte.
*Dioscorea amaranthoides* var. *ulei* R.Knuth, Notizbl.
Bot. Gart. Berlin-Dahlem 7: 215 (1917).

**Dioscorea pentaphylla** L., Sp. Pl.: 1032 (1753).
*Botryosicyos pentaphyllus* (L.) Hochst., Flora 27(Bes.
Beil.): 3 (1844).

Trop. & Subtrop. Asia to N. Australia. 36 CHC CHH
CHS CHT 38 NNS TAI 40 ASS BAN EHM IND
NEP SRL WHM 41 LAO MYA THA VIE 42 BOR
JAW LSI MLY MOL PHI SUL SUM 43 NWG 50
QLD (60) fij nue ton (61) mrq sci 62 CRL (63) haw
(7). Cl. tuber geophyte.
*Dioscorea triphylla* L., Sp. Pl.: 1032 (1753). *Hamatris
triphylla* (L.) Salisb., Gen. Pl.: 12 (1866).
*Dioscorea digitata* Mill., Gard. Dict. ed. 8: 6 (1768).
*Dioscorea spinosa* Burm., Fl. Malab.: 5 (1769).
*Ubium quadrifarium* J.F.Gmel., Syst. Nat. 2: 839
(1791).
*Ubium scandens* J.St.-Hil., Expos. Fam. Nat. 1: 106
(1805).
*Dioscorea kleiniana* Kunth, Enum. Pl. 5: 394 (1850).
*Dioscorea jacquemontii* Hook.f., Fl. Brit. India 6: 290
(1892). *Dioscorea pentaphylla* var. *jacquemontii*
(Hook.f.) Prain & Burkill, J. Proc. Asiat. Soc.
Bengal 10(1): 23 (1914).
*Dioscorea pentaphylla* var. *cardonii* Prain & Burkill, J.
Proc. Asiat. Soc. Bengal 10(1): 23 (1914).
*Dioscorea pentaphylla* var. *communis* Prain & Burkill,
J. Proc. Asiat. Soc. Bengal 10(1): 23 (1914).
*Dioscorea pentaphylla* var. *hortorum* Prain & Burkill,
J. Proc. Asiat. Soc. Bengal 10(1): 23 (1914).
*Dioscorea pentaphylla* var. *kussok* Prain & Burkill, J.
Proc. Asiat. Soc. Bengal 10(1): 23 (1914).
*Dioscorea pentaphylla* var. *linnaei* Prain & Burkill, J.
Proc. Asiat. Soc. Bengal 10(1): 23 (1914).
*Dioscorea pentaphylla* var. *malaica* Prain & Burkill, J.
Proc. Asiat. Soc. Bengal 10(1): 23 (1914).
*Dioscorea pentaphylla* var. *rheedei* Prain & Burkill, J.
Proc. Asiat. Soc. Bengal 10(1): 23 (1914).
*Dioscorea pentaphylla* var. *simplicifolia* Prain &
Burkill, J. Proc. Asiat. Soc. Bengal 10(1): 23
(1914).
*Dioscorea pentaphylla* var. *suli* Prain & Burkill, J.
Proc. Asiat. Soc. Bengal 10(1): 23 (1914).
*Dioscorea pentaphylla* var. *thwaitesii* Prain & Burkill,
J. Proc. Asiat. Soc. Bengal 10(1): 23 (1914).
*Dioscorea globifera* R.Knuth in H.G.A.Engler (ed.),
Pflanzenr., IV, 43: 149 (1924).
*Dioscorea pentaphylla* var. *papuana* Burkill, Gard.
Bull. Straits Settlem. 3: 258 (1924).
*Dioscorea pentaphylla* var. *unifoliata* R.Knuth in
H.G.A.Engler (ed.), Pflanzenr., IV, 43: 146 (1924).
*Dioscorea codonopsidifolia* Kamik., Trans. Nat. Hist.
Soc. Taiwan 25: 115 (1935).
*Dioscorea changjiangensis* F.W.Xing & Z.X.Li, Acta
Bot. Austro Sin. 10: 19 (1995).

**Dioscorea peperoides** Prain & Burkill, Leafl. Philipp.
Bot. 5: 1597 (1913).
Vietnam, Philippines. 41 VIE 42 PHI. Cl. tuber
geophyte.

var. **angulata** Prain & Burkill, J. Proc. Asiat. Soc.
Bengal 10: 28 (1914).
Vietnam. 41 VIE.

var. **peperoides**
Philippines. 42 PHI. Cl. tuber geophyte.
*Dioscorea peperoides* var. *sagittifolia* Prain &
Burkill, J. Proc. Asiat. Soc. Bengal 10: 28
(1914).
*Dioscorea peperoides* var. *vera* Prain & Burkill, J.
Proc. Asiat. Soc. Bengal 10: 28 (1914).

**Dioscorea perdicum** Taub., Bot. Jahrb. Syst. 15(34): 13
(1892).
Brazil (Rio de Janeiro). 84 BZL. Cl. tuber geophyte.

*Dioscorea perenensis* R.Knuth, Repert. Spec. Nov. Regni Veg. 29: 92 (1931).
Peru. 83 PER. Cl. tuber geophyte.

*Dioscorea perpilosa* H.Perrier, Notul. Syst. (Paris) 12: 206 (1946).
NW. Madagascar. 29 MDG. Cl. tuber geophyte.

*Dioscorea persimilis* Prain & Burkill, J. Proc. Asiat. Soc. Bengal 4: 454 (1908).
S. China to Vietnam. 36 CHH CHS 38 TAI 41 VIE. Cl. tuber geophyte.
  *Dioscorea raishaensis* Hayata, Icon. Pl. Formosan. 10: 44 (1921).
  *Dioscorea persimilis* var. *pubescens* C.T.Ting & M.C.Chang, Acta Phytotax. Sin. 20: 205 (1982).

*Dioscorea petelotii* Prain & Burkill, Bull. Misc. Inform. Kew 1933: 240 (1933).
N. Thailand to N. Vietnam. 41 THA VIE. Cl. tuber geophyte.

*Dioscorea philippiana* Uline ex R.Knuth, Notizbl. Bot. Gart. Berlin-Dahlem 7: 193 (1917).
Chile. 85 CLC. Cl. tuber geophyte.

*Dioscorea piauhyensis* R.Knuth in H.G.A.Engler (ed.), Pflanzenr., IV, 43: 64 (1924).
NE. Brazil. 84 BZE. Cl. tuber geophyte.

*Dioscorea pierrei* Prain & Burkill, J. Proc. Asiat. Soc. Bengal 10: 22 (1914).
Indo-China. 41 CBD THA VIE 42 MLY?. Cl. tuber geophyte.

*Dioscorea pilcomayensis* Hauman, Anales Mus. Nac. Hist. Nat. Buenos Aires 27: 502 (1916).
Bolivia to N. Argentina. 83 BOL 85 AGE AGW PAR. Cl. tuber geophyte.
  *\*Dioscorea pedicellata* Morong, Ann. New York Acad. Sci. 7: 240 (1893), nom. illeg. *Dioscorea pellegrinii* Hassl. ex R.Knuth in H.G.A.Engler (ed.), Pflanzenr., IV, 43: 108 (1924), nom. illeg.
  *Dioscorea tamifolia* Chodat & Hassl., Bull. Herb. Boissier, II, 3: 1110 (1903), nom. illeg.

*Dioscorea pilgeriana* R.Knuth, Repert. Spec. Nov. Regni Veg. 42: 177 (1937).
Brazil (Minas Gerais). 84 BZL. Cl. tuber geophyte.

*Dioscorea pilosiuscula* Bertero ex Spreng., Syst. Veg. 2: 152 (1825). *Helmia pilosiuscula* (Bertero ex Spreng.) Kunth, Enum. Pl. 5: 434 (1850).
Trop. America. 79 MXT 80 COS GUA HON NIC PAN 81 DOM HAI JAM LEE PUE TRT WIN 82 FRG GUY SUR VEN 83 BOL CLM ECU PER 84 BZC BZL. Cl. tuber geophyte.
  *Dioscorea sapindioides* C.Presl, Reliq. Haenk. 1: 133 (1827).
  *Dioscorea cuspidata* Balb. ex Kunth, Enum. Pl. 5: 434 (1850).
  *Helmia schomburgkiana* Kunth, Enum. Pl. 5: 424 (1850). *Dioscorea schomburgkiana* (Kunth) Hochr., Bull. New York Bot. Gard. 6: 268 (1916).
  *Dioscorea costaricensis* R.Knuth, Notizbl. Bot. Gart. Berlin-Dahlem 7: 189 (1917).
  *Dioscorea lindmanii* Uline ex R.Knuth, Notizbl. Bot. Gart. Berlin-Dahlem 7: 191 (1917).
  *Dioscorea pilosiuscula* var. *panamensis* R.Knuth in H.G.A.Engler (ed.), Pflanzenr., IV, 43: 66 (1924).
  *Dioscorea lanosa* Gleason, Bull. Torrey Bot. Club 52: 181 (1925).

*Dioscorea pinedensis* R.Knuth, Repert. Spec. Nov. Regni Veg. 29: 95 (1931).
Peru. 83 PER. Cl. tuber geophyte.

*Dioscorea piperifolia* Humb. & Bonpl. ex Willd., Sp. Pl. 4: 795 (1806).
S. Trop. America. 82 GUY VEN 83 CLM ECU PER 84 BZC BZE BZL BZN BZS 85 AGE PAR. Cl. tuber geophyte.
  *Dioscorea conferta* Vell., Fl. Flumin. 10: t. 122 (1831).
  *Dioscorea maynensis* Kunth, Enum. Pl. 5: 357 (1850).
  *Sismondaea dioscoreoides* Delponte, Mem. Reale Accad. Sci. Torino, II, 14: 394 (1854).
  *Dioscorea piperifolia* var. *obtusifolia* Chodat & Hassl., Bull. Herb. Boissier, II, 3: 1110 (1903).
  *Dioscorea piperifolia* var. *apiculata* Uline, Notizbl. Bot. Gart. Berlin-Dahlem 7: 196 (1917).

*Dioscorea piscatorum* Prain & Burkill, Gard. Bull. Straits Settlem. 3: 123 (1924).
W. Malesia. 42 BOR MLY SUM. Cl. tuber geophyte.
  *Dioscorea borneensis* R.Knuth in H.G.A.Engler (ed.), Pflanzenr., IV, 43: 188 (1924).

*Dioscorea pittieri* R.Knuth, Repert. Spec. Nov. Regni Veg. 28: 96 (1930).
Venezuela. 82 VEN. Cl. tuber geophyte.

*Dioscorea planistipulosa* Uline ex R.Knuth, Notizbl. Bot. Gart. Berlin-Dahlem 7: 204 (1917).
E. Brazil. 84 BZE BZL. Cl. tuber geophyte.

  var. *glaziovii* R.Knuth in H.G.A.Engler (ed.), Pflanzenr., IV, 43: 169 (1924).
  SE. Brazil. 84 BZL.

  var. *planistipulosa*
  NE. Brazil. 84 BZE. Cl. tuber geophyte.

*Dioscorea plantaginifolia* R.Knuth, Notizbl. Bot. Gart. Berlin-Dahlem 7: 211 (1917).
Brazil (Rio de Janeiro). 84 BZL. Cl. tuber geophyte.

*Dioscorea platycarpa* Prain & Burkill, Bull. Misc. Inform. Kew 1925: 65 (1925).
Jawa. 42 JAW. Cl. tuber geophyte.

*Dioscorea platycolpota* Uline ex B.L.Rob., Proc. Amer. Acad. Arts 36: 471 (1900).
C. & SW. Mexico. 79 MXC MXS. Cl. tuber geophyte.

*Dioscorea plumifera* C.B.Rob., Proc. Amer. Acad. Arts 29: 324 (1894).
Mexico. 79 MXC MXN MXS. Cl. tuber geophyte.

*Dioscorea pohlii* Griseb. in C.F.P.von Martius & auct. suc. (eds.), Fl. Bras. 3(1): 35 (1842).
Brazil. 84 BZC BZE BZL. Cl. tuber geophyte.

  var. *luschnathiana* (Kunth) Uline ex R.Knuth, Notizbl. Bot. Gart. Berlin-Dahlem 7: 213 (1917).
  E. Brazil. 84 BZE BZL.
    *\*Dioscorea luschnathiana* Kunth, Enum. Pl. 5: 364 (1850).

  var. *pohlii*
  WC. Brazil. 84 BZC. Cl. tuber geophyte.

*Dioscorea poilanei* Prain & Burkill, Bull. Misc. Inform. Kew 1933: 240 (1933).
Hainan to Pen. Malaysia. 36 CHH 41 CBD LAO THA VIE 42 MLY. Cl. rhizome geophyte.

*Dioscorea polyanthes* (F.Phil.) Caddick & Wilkin, Taxon 51: 112 (2002).
C. Chile. 85 CLC. Cl. tuber geophyte.
  *\*Epipetrum polyanthes* F.Phil., Anales Univ. Chile 93: 22 (1896).

*Dioscorea polyclados* Hook.f., Fl. Brit. India 6: 294 (1892).
Indo-China to W. Malesia. 41 VIE 42 BOR JAW MLY SUM. Cl. tuber geophyte.
*Dioscorea polyclados* var. *oblongifolia* Uline ex R.Knuth in H.G.A.Engler (ed.), Pflanzenr., IV, 43: 275 (1924).

*Dioscorea polygonoides* Humb. & Bonpl. ex Willd., Sp. Pl. 4: 795 (1806).
S. Mexico to Trop. America. 79 MXS MXT 80 COS ELS GUA NIC PAN 81 CUB DOM HAI JAM LEE PUE TRT WIN 82 FRG GUY SUR VEN 83 BOL CLM PER 84 BZC BZE BZL BZN BZS 85 AGE AGW. Cl. tuber geophyte.
*Dioscorea lutea* G.Mey., Prim. Fl. Esseq.: 282 (1818).
*Dioscorea martinicensis* Spreng., Neue Entdeck. Pflanzenk. 3: 17 (1822). *Dioscorea polygonoides* var. *martinicensis* (Spreng.) R.Knuth in H.G.A. Engler (ed.), Pflanzenr., IV, 43: 217 (1924).
*Dioscorea piperifolia* Klotzsch ex Kunth, Enum. Pl. 3: 354 (1841), nom. illeg.
*Dioscorea piperifolia* Griseb. in C.F.P.von Martius & auct. suc. (eds.), Fl. Bras. 3(1): 42 (1842), nom. illeg.
*Dioscorea altissima* Sieber ex C.Presl, Abh. Königl. Böhm. Ges. Wiss., V, 3: 546 (1845), nom. illeg.
*Dioscorea multiflora* C.Presl, Abh. Königl. Böhm. Ges. Wiss., V, 3: 546 (1845), nom. illeg.
*Dioscorea kegeliana* Griseb., Linnaea 21: 279 (1848).
*Dioscorea sieberi* Kunth, Abh. Königl. Akad. Wiss. Berlin 1848: 58 (1848). *Dioscorea polygonoides* var. *sieberi* (Kunth) Uline, Notizbl. Bot. Gart. Berlin-Dahlem 7: 208 (1917).
*Dioscorea caracasana* Kunth, Enum. Pl. 5: 354 (1850).
*Dioscorea polygonoides* var. *aperta* R.Knuth, Notizbl. Bot. Gart. Berlin-Dahlem 7: 208 (1917).
*Dioscorea polygonoides* var. *scorpioidea* Uline, Notizbl. Bot. Gart. Berlin-Dahlem 7: 208 (1917).

*Dioscorea polystachya* Turcz., Bull. Soc. Imp. Naturalistes Moscou 7: 158 (1837).
C. China to Temp. E. Asia. (27) 31 KUR 36 CHC CHN CHS 38 JAP KOR NNS TAI (74) mso (75) cnt mas nwj ohi pen wva (78) ala ark fla geo kty mry msi nca sca ten vrg wdc. Cl. tuber geophyte.
*Dioscorea batatas* Decne., Rev. Hort., IV, 3: 243 (1854).
*Dioscorea decaisneana* Carrière, Rev. Hort. 1865: 111 (1865).
*Dioscorea doryphora* Hance, Ann. Sci. Nat., Bot., V, 5: 244 (1866).
*Dioscorea swinhoei* Rolfe, J. Bot. 20: 359 (1882).
*Dioscorea rosthornii* Diels, Bot. Jahrb. Syst. 29: 261 (1900).
*Dioscorea batatas* f. *clavata* Makino in Y.Iinuma, Somoku-Dzusetsu, ed. 3, 4: 1326 (1912).
*Dioscorea batatas* f. *daikok* Makino in Y.Iinuma, Somoku-Dzusetsu, ed. 3, 4: 1326 (1912).
*Dioscorea batatas* f. *flabellata* Makino in Y.Iinuma, Somoku-Dzusetsu, ed. 3, 4: 1326 (1912).
*Dioscorea batatas* f. *rakuda* Makino in Y.Iinuma, Somoku-Dzusetsu, ed. 3, 4: 1326 (1912).
*Dioscorea batatas* f. *tsukune* Makino in Y.Iinuma, Somoku-Dzusetsu, ed. 3, 4: 1326 (1912).
*Dioscorea cayennensis* var. *pseudobatatas* Hauman, Anales Mus. Nac. Hist. Nat. Buenos Aires 27: 488 (1915). *Dioscorea pseudobatatas* (Hauman) Herter, Fl. Illustr. Urug. 1: 235 (1943).
*Dioscorea potaninii* Prain & Burkill, Bull. Misc. Inform. Kew 1933: 243 (1933).

*Dioscorea pomeroonensis* R.Knuth, Notizbl. Bot. Gart. Berlin-Dahlem 7: 217 (1917).
Guyana. 82 GUY. Cl. tuber geophyte.

*Dioscorea potarensis* R.Knuth in H.G.A.Engler (ed.), Pflanzenr., IV, 43: 240 (1924).
Guyana. 82 GUY. Cl. tuber geophyte.

*Dioscorea praehensilis* Benth. in W.J.Hooker, Niger Fl.: 536 (1849). *Dioscorea cayennensis* var. *praehensilis* (Benth.) A.Chev., Bull. Mus. Natl. Hist. Nat., II, 8: 537 (1936).
Trop. Africa. 22 BEN BKN GHA IVO LBR NGA SIE TOG 23 BUR GAB RWA ZAI 24 CHA 25 KEN TAN UGA 26 ANG MLW ZAM. Cl. tuber geophyte.
*Dioscorea odoratissima* Pax, Bot. Jahrb. Syst. 15: 146 (1892).
*Dioscorea angustiflora* Rendle in W.P.Hiern, Cat. Afr. Pl. 2: 39 (1899).
*Dioscorea costermansiana* De Wild. & T.Durand, Bull. Herb. Boissier, II, 1: 52 (1900).
*Dioscorea liebrechtsiana* De Wild. & T.Durand, Bull. Herb. Boissier, II, 1: 53 (1900).

*Dioscorea prainiana* R.Knuth, Bull. Misc. Inform. Kew 1925: 286 (1925).
W. & C. Malesia. 42 MLY SUL SUM. Cl. tuber geophyte.
*\*Dioscorea deflexa* Hook.f., Fl. Brit. India 6: 293 (1892), nom. illeg.
*Dioscorea maliliensis* R.Knuth, Repert. Spec. Nov. Regni Veg. 36: 127 (1934).

*Dioscorea prazeri* Prain & Burkill, J. Asiat. Soc. Bengal, Pt. 2, Nat. Hist. 73(Suppl.): 2 (1904).
C. Himalaya to Pen. Malaysia. 40 ASS BAN EHM NEP 41 MYA THA VIE 42 MLY. Rhizome geophyte.
*Dioscorea deltoidea* var. *sikkimensis* Prain, Bengal Pl. 2: 1066 (1903).
*Dioscorea sikkimensis* Prain & Burkill, J. Asiat. Soc. Bengal, Pt. 2, Nat. Hist. 73(Suppl.): 3 (1904).
*Dioscorea clarkei* Prain & Burkill, J. Proc. Asiat. Soc. Bengal 10: 15 (1914).

*Dioscorea preslii* Steud., Nomencl. Bot., ed. 2, 1: 511 (1840).
SW. Mexico. 79 MXS. Cl. tuber geophyte.
*\*Dioscorea hastata* C.Presl, Reliq. Haenk. 1: 133 (1827), nom. illeg.

*Dioscorea preussii* Pax, Bot. Jahrb. Syst. 15: 147 (1892).
Trop. Africa. 22 BEN GHA GUI IVO NGA SEN SIE TOG 23 CAF CMN GAB ZAI 24 SUD 25 TAN UGA 26 ANG MLW MOZ ZAM ZIM. Cl. tuber geophyte.

subsp. *hylophila* (Harms) Wilkin, Kew Bull. 56: 394 (2001).
Tanzania to S. Trop. Africa. 25 TAN 26 MLW ZAM ZIM. Cl. tuber geophyte.
*\*Dioscorea hylophila* Harms in H.G.A.Engler, Pflanzenw. Ost-Afrikas, C: 146 (1895).

subsp. *preussii*
Trop. Africa. 22 BEN GHA GUI IVO NGA SEN SIE TOG 23 CAF CMN GAB ZAI 24 SUD 25 TAN? UGA 26 ANG MOZ. Cl. tuber geophyte.
*Dioscorea andongensis* Rendle in W.P.Hiern, Cat. Afr. Pl. 2: 35 (1899).
*Dioscorea pterocaulon* De Wild. & T.Durand, Ann. Mus. Congo Belge, Bot., II, 1(1): 58 (1899).

*Dioscorea thonneri* De Wild. & T.Durand, Ann. Mus. Congo Belge, Bot. 1: 109 (1900).

*Dioscorea malchairii* De Wild., Ann. Mus. Congo Belge, Bot., V, 3: 365 (1912).

*Dioscorea longispicata* De Wild., Bull. Jard. Bot. État 4: 323 (1913).

*Dioscorea chevalieri* De Wild., Bull. Jard. Bot. État 4: 316 (1914).

*Dioscorea dawei* De Wild., Bull. Jard. Bot. État 4: 317 (1914).

**Dioscorea pringlei** C.B.Rob., Proc. Amer. Acad. Arts 29: 323 (1894).
C. & SW. Mexico. 79 MXC MXS. Cl. tuber geophyte.

**Dioscorea proteiformis** H.Perrier, Notul. Syst. (Paris) 12: 199 (1946).
E. Madagascar. 29 MDG. Cl. tuber geophyte.

**Dioscorea psammophila** R.Knuth, Repert. Spec. Nov. Regni Veg. 38: 120 (1935).
Brazil (Bahia). 84 BZE. Cl. tuber geophyte.

**Dioscorea pseudomacrocapsa** G.M.Barroso, E.F.Guim. & Sucre, Revista Brasil. Biol. 31: 309 (1971).
Brazil. 84+. Cl. tuber geophyte.

**Dioscorea pseudonitens** Prain & Burkill, Bull. Misc. Inform. Kew 1927: 231 (1927).
N. Thailand. 41 THA. Cl. tuber geophyte.

**Dioscorea pseudorajanioides** R.Knuth, Repert. Spec. Nov. Regni Veg. 22: 347 (1926).
Colombia. 83 CLM. Cl. tuber geophyte.
*Dioscorea truncata Rusby, Descr. S. Amer. Pl.: 6 (1920), nom. illeg.

**Dioscorea pseudotomentosa** Prain & Burkill, Bull. Misc. Inform. Kew 1927: 234 (1927).
Thailand. 41 THA. Cl. tuber geophyte.

**Dioscorea pteropoda** Boivin ex H.Perrier, Mém. Soc. Linn. Normandie, Bot. 1(1): 17 (1928).
N. Madagascar. 29 MDG. Cl. tuber geophyte.

**Dioscorea pubera** Blume, Enum. Pl. Javae 1: 21 (1827).
C. Himalaya to W. Malesia. 40 ASS BAN EHM IND NEP 41 MYA 42 JAW SUM. Cl. tuber geophyte.
*Dioscorea combilium* Buch.-Ham. in N.Wallich, Numer. List: 5103A (1830), nom. inval.
*Dioscorea anguina* Roxb., Fl. Ind. ed. 1832, 3: 803 (1832).
*Dioscorea cornifolia* Kunth, Enum. Pl. 5: 385 (1850).

**Dioscorea pubescens** Poir. in J.B.A.M.de Lamarck, Encycl., Suppl. 3: 137 (1813).
French Guiana. 82 FRG. Cl. tuber geophyte.

**Dioscorea pumicicola** Uline, Proc. Amer. Acad. Arts 35: 322 (1900).
C. Mexico. 79 MXC. Cl. tuber geophyte.

**Dioscorea pumilio** Griseb., Vidensk. Meddel. Naturhist. Foren. Kjøbenhavn 1875: 162 (1875).
Brazil (Rio de Janeiro). 84 BZL. Tuber geophyte.

**Dioscorea puncticulata** R.Knuth in H.G.A.Engler (ed.), Pflanzenr., IV, 43: 55 (1924).
S. Brazil. 84 BZS. Cl. tuber geophyte.

**Dioscorea purdiei** R.Knuth in H.G.A.Engler (ed.), Pflanzenr., IV, 43: 100 (1924).
W. South America. 83 CLM PER. Cl. tuber geophyte.

**Dioscorea putisensis** R.Knuth, Repert. Spec. Nov. Regni Veg. 28: 82 (1930).
Peru. 83 PER. Cl. tuber geophyte.

**Dioscorea putumayensis** R.Knuth, Repert. Spec. Nov. Regni Veg. 30: 160 (1932).
Colombia. 83 CLM. Cl. tuber geophyte.

**Dioscorea pynaertii** De Wild., Ann. Mus. Congo Belge, Bot., V, 3: 366 (1912).
WC. Trop. Africa. 23 GAB ZAI. Cl. tuber geophyte.

**Dioscorea pyrenaica** Bubani & Bordère ex Gren., Bull. Soc. Bot. France 13: 382 (1866).
C. Pyrenees. 12 FRA SPA. Tuber geophyte.
*Borderea pyrenaica* Miégev., Bull. Soc. Bot. France 13: 374 (1866).

**Dioscorea pyrifolia** Kunth, Enum. Pl. 5: 384 (1850).
E. Himalaya to W. Malesia. 40 ASS EHM 41 THA VIE 42 BOR JAW MLY SUM. Cl. tuber geophyte.

var. **ferruginea** Prain & Burkill, J. Proc. Asiat. Soc. Bengal 10: 33 (1914).
E. Himalaya to Assam. 40 ASS EHM. Cl. tuber geophyte.

var. **pyrifolia**
Pen. Thailand to W. Malesia. 41 THA VIE 42 BOR JAW MLY SUM. Cl. tuber geophyte.
*Dioscorea zollingeriana* Kunth, Enum. Pl. 5: 384 (1850).
*Dioscorea diepenhorstiana* Miq., Fl. Ned. Ind., Eerste Bijv.: 611 (1861).
*Dioscorea ferruginea* Thunb. ex Prain & Burkill, J. Proc. Asiat. Soc. Bengal 10: 33 (1914), nom. inval.
*Dioscorea preangeriana* Uline ex R.Knuth in H.G.A. Engler (ed.), Pflanzenr., IV, 43: 269 (1924).
*Dioscorea sandakanensis* R.Knuth, Repert. Spec. Nov. Regni Veg. 36: 127 (1934).

**Dioscorea quartiniana** A.Rich., Tent. Fl. Abyss. 2: 316 (1850).
Trop. & S. Africa, Madagascar. 22 BEN GAM GHA IVO NGA SIE 23 BUR RWA ZAI 24 CHA ERI ETH SUD 25 KEN TAN UGA 26 MLW MOZ ZAM ZIM 27 BOT CPV NAM NAT SWZ TVL 29 MDG. Cl. tuber geophyte.
*Dioscorea vespertilio* Benth. in W.J.Hooker, Niger Fl.: 538 (1849).
*Dioscorea crinita* Hook.f., Bot. Mag. 111: t. 6804 (1885).
*Dioscorea beccariana* Martelli, Fl. Bogos.: 83 (1886).
*Dioscorea cryptantha* Baker, J. Linn. Soc., Bot. 22: 528 (1887). *Dioscorea quartiniana* var. *cryptantha* (Baker) Burkill, Bull. Jard. Bot. État 15: 365 (1939).
*Dioscorea forbesii* Baker, J. Bot. 27: 2 (1889). *Dioscorea quartiniana* var. *forbesii* (Baker) Burkill, Bull. Jard. Bot. État 15: 365 (1939).
*Dioscorea phaseoloides* Pax, Bot. Jahrb. Syst. 15: 149 (1892). *Dioscorea quartiniana* var. *phaseoloides* (Pax) Burkill, Bull. Jard. Bot. État 15: 365 (1939).
*Dioscorea quartiniana* var. *hochstetteri* Engl., Hochgebirgsfl. Afrika: 172 (1892).
*Dioscorea quartiniana* var. *pentadactyla* Pax, Bot. Jahrb. Syst. 15: 148 (1892). *Dioscorea pentadactyla* (Pax) Welw. in W.P.Hiern, Cat. Afr. Pl. 2(1): 41 (1899).
*Dioscorea schweinfurthiana* Pax, Bot. Jahrb. Syst. 15: 149 (1892). *Dioscorea quartiniana* var. *schweinfurthiana* (Pax) Burkill, Bull. Jard. Bot. État 15: 365 (1939).
*Dioscorea holstii* Harms in H.G.A.Engler, Pflanzenw. Ost-Afrikas, C: 147 (1895). *Dioscorea quartiniana* var. *holstii* (Harms) Burkill, Bull. Jard. Bot. État 15: 365 (1939).

*Dioscorea stuhlmannii* Harms in H.G.A.Engler, Pflanzenw. Ost-Afrikas, C: 146 (1895). *Dioscorea quartiniana* var. *stuhlmannii* (Harms) Burkill, Bull. Jard. Bot. État 15: 365 (1939).
*Dioscorea dinteri* Schinz, Mém. Herb. Boissier 20: 11 (1900). *Dioscorea quartiniana* var. *dinteri* (Schinz) Burkill, Bull. Jard. Bot. État 15: 365 (1939).
*Dioscorea apiculata* De Wild., Ann. Mus. Congo Belge, Bot., IV, 1: 14 (1902). *Dioscorea quartiniana* var. *apiculata* (De Wild.) Burkill, Bull. Jard. Bot. État 15: 365 (1939).
*Dioscorea verdickii* De Wild., Ann. Mus. Congo Belge, Bot., IV, 1: 15 (1902).
*Dioscorea anchiatasi* Harms in H.G.A.Engler & C.G.O. Drude, Veg. Erde 9(2): 362 (1908), nom. inval.
*Dioscorea quartiniana* var. *subpedata* Chiov., Ann. Bot. (Rome) 9: 142 (1911).
*Dioscorea anchietae* Harms ex R.Knuth in H.G.A.Engler (ed.), Pflanzenr., IV, 43: 155 (1924).
*Dioscorea excisa* R.Knuth in H.G.A.Engler (ed.), Pflanzenr., IV, 43: 155 (1924). *Dioscorea quartiniana* var. *excisa* (R.Knuth) Burkill, Bull. Jard. Bot. État 15: 365 (1939).
*Dioscorea schliebenii* R.Knuth, Notizbl. Bot. Gart. Berlin-Dahlem 11: 659 (1932).
*Dioscorea ulugurensis* R.Knuth, Notizbl. Bot. Gart. Berlin-Dahlem 11: 1060 (1934).
*Dioscorea angolensis* R.Knuth, Repert. Spec. Nov. Regni Veg. 38: 119 (1935).
*Dioscorea gossweileri* R.Knuth, Repert. Spec. Nov. Regni Veg. 38: 120 (1935).
*Dioscorea peteri* R.Knuth, Repert. Spec. Nov. Regni Veg. 42: 161 (1937).
*Dioscorea quartiniana* var. *latifolia* R.Knuth, Repert. Spec. Nov. Regni Veg. 42: 161 (1937).
*Dioscorea quartiniana* var. *vestita* R.Knuth, Repert. Spec. Nov. Regni Veg. 42: 161 (1937).
*Dioscorea quartiniana* var. *schliebenii* Burkill, Bull. Jard. Bot. État 15: 365 (1939).

**Dioscorea quinquelobata** Thunb. in J.A.Murray, Syst. Veg. ed. 14: 889 (1784).
China to Temp. E. Asia. 36 CHS 38 JAP KOR NNS. Cl. rhizome geophyte.
*Dioscorea quinqueloba* Thunb., Fl. Jap.: 150 (1784).

**Dioscorea quispicanchensis** R.Knuth, Repert. Spec. Nov. Regni Veg. 39: 93 (1931).
Peru. 83 PER. Cl. tuber geophyte.

**Dioscorea racemosa** (Klotzsch) Uline, Bot. Jahrb. Syst. 22: 430 (1896).
SE. Mexico to Panama. 79 MXT 80 COS PAN. Cl. tuber geophyte.
*\*Helmia racemosa* Klotzsch, Allg. Gartenzeitung 19: 393 (1851).
*Dioscorea borealis* C.V.Morton, J. Wash. Acad. Sci. 27: 304 (1937).
*Dioscorea coxii* Matuda, Bol. Soc. Bot. México 21: 6 (1957).

**Dioscorea regnellii** Uline ex R.Knuth, Notizbl. Bot. Gart. Berlin-Dahlem 7: 214 (1917).
Brazil (São Paulo). 84 BZL. Cl. tuber geophyte.

**Dioscorea remota** C.V.Morton, J. Wash. Acad. Sci. 27: 304 (1937).
Costa Rica to Panama. 80 COS PAN. Cl. tuber geophyte.

**Dioscorea remotiflora** Kunth, Enum. Pl. 5: 409 (1850).
Mexico. 79 MXC. Cl. tuber geophyte.

*Dioscorea laxiflora* Schltdl., Linnaea 17: 606 (1843), nom. illeg.
*Helmia ehrenbergiana* Kunth, Enum. Pl. 5: 433 (1850).
*Dioscorea sparsiflora* Hemsl., Biol. Cent.-Amer., Bot. 3: 360 (1884). *Dioscorea remotiflora* var. *sparsiflora* (Hemsl.) Uline, Bot. Jahrb. Syst. 22: 422 (1897).
*Dioscorea remotiflora* var. *maculata* Uline, Bot. Jahrb. Syst. 22: 422 (1897).
*Dioscorea remotiflora* var. *palmeri* Uline, Bot. Jahrb. Syst. 22: 422 (1897).

**Dioscorea reticulata** Gay, Fl. Chil. 6: 61 (1854). *Dioscorea brachybotrya* var. *reticulata* (Gay) Uline ex R.Knuth, Notizbl. Bot. Gart. Berlin-Dahlem 7: 206 (1917).
C. Chile, Argentina (Neuquén). 85 AGS CLC. Cl. tuber geophyte.
*Dioscorea pedatifida* Phil., Anales Univ. Chile 93: 18 (1896).
*Dioscorea thermarum* Phil., Anales Univ. Chile 93: 17 (1896).

**Dioscorea retusa** Mast., Gard. Chron. 1870: 1149 (1870).
S. Africa. 27 CPP NAT SWZ TVL. Cl. tuber geophyte.
*Dioscorea tysonii* Baker, J. Bot. 27: 2 (1889).
*Dioscorea microcuspis* Baker in W.H.Harvey & auct. suc. (eds.), Fl. Cap. 6: 250 (1896).

**Dioscorea reversiflora** Uline, Bot. Jahrb. Syst. 22: 426 (1896).
Mexico (Sinaloa, Nayarit). 79 MXN MXS. Cl. tuber geophyte.

**Dioscorea ridleyi** Prain & Burkill, J. Proc. Asiat. Soc. Bengal 10: 12 (1914).
Borneo (Sarawak). 42 BOR. Cl. tuber geophyte.

**Dioscorea riedelii** R.Knuth, Notizbl. Bot. Gart. Berlin-Dahlem 7: 213 (1917).
SE. Brazil. 84 BZL. Cl. tuber geophyte.

**Dioscorea rigida** R.Knuth, Repert. Spec. Nov. Regni Veg. 21: 79 (1925).
W. Cuba. 81 CUB. Cl. tuber geophyte.

**Dioscorea rimbachii** R.Knuth, Repert. Spec. Nov. Regni Veg. 22: 344 (1926).
Ecuador. 83 ECU. Cl. tuber geophyte.

**Dioscorea rockii** Prain & Burkill, Bull. Misc. Inform. Kew 1927: 229 (1927).
N. Thailand. 41 THA. Cl. rhizome geophyte.

**Dioscorea rosei** R.Knuth in H.G.A.Engler (ed.), Pflanzenr., IV, 43: 229 (1924).
Ecuador. 83 ECU. Cl. tuber geophyte.

**Dioscorea rumicoides** Griseb. in C.F.P.von Martius & auct. suc. (eds.), Fl. Bras. 3(1): 42 (1842).
Brazil to Peru. 83 BOL PER 84 BZC BZL BZS. Cl. tuber geophyte.
*Dioscorea rumicoides* var. *longibracteata* R.Knuth, Repert. Spec. Nov. Regni Veg. 28: 86 (1930).

**Dioscorea rupicola** Kunth, Enum. Pl. 5: 378 (1850).
S. Africa. 27 CPP NAT SWZ TVL. Cl. tuber geophyte.

**Dioscorea rusbyi** Uline in H.G.A.Engler & K.A.E.Prantl, Nat. Pflanzenfam., Nachtr. 1: 86 (1897).
Bolivia. 83 BOL. Cl. tuber geophyte.

**Dioscorea sabarensis** R.Knuth in H.G.A.Engler (ed.), Pflanzenr., IV, 43: 241 (1924).
Brazil (Minas Gerais). 84 BZL. Cl. tuber geophyte.

*Dioscorea sagittata* Poir. in J.B.A.M.de Lamarck, Encycl., Suppl. 3: 139 (1813).
French Guiana. 82 FRG. Cl. tuber geophyte.

*Dioscorea sagittifolia* Pax, Bot. Jahrb. Syst. 15: 147 (1892).
W. Trop. Africa to Uganda. 22 BEN BKN GUI IVO LBR MLI SEN SIE TOG 23 CAF CMN CON ZAI 24 CHA SUD 25 UGA. Cl. tuber geophyte.

var. **lecardii** (De Wild.) Nkounkou, Belgian J. Bot. 126: 62 (1993).
W. Trop. Africa to Uganda. 22 BEN GUI IVO MLI SEN SIE TOG 23 CAF CMN CON ZAI 24 CHA 25 UGA. Cl. tuber geophyte.
*\*Dioscorea lecardii* De Wild., Ann. Mus. Congo Belge, Bot., V, 1: 19 (1903).
*Dioscorea zara* Baudon, Ann. Inst. Bot.-Géol. Colon. Marseille, III, 1: 237 (1913).
*Dioscorea mildbraedii* R.Knuth in H.G.A.Engler (ed.), Pflanzenr., IV, 43: 295 (1924).

var. **sagittifolia**
W. Trop. Africa to Sudan. 22 BKN IVO LBR SEN 23 CAF CMN CON ZAI 24 SUD. Cl. tuber geophyte.

*Dioscorea salicifolia* Blume, Enum. Pl. Javae 1: 23 (1827).
W. Malesia. 42 BOR JAW SUM. Cl. tuber geophyte.
*Dioscorea gracillima* Ridl., Bot. Jahrb. Syst. 44: 528 (1910), nom. illeg. *Dioscorea sarawakensis* R.Knuth in H.G.A.Engler (ed.), Pflanzenr., IV, 43: 291 (1924).

*Dioscorea salvadorensis* Standl., J. Wash. Acad. Sci. 13: 365 (1923).
El Salvador. 80 ELS. Cl. tuber geophyte.

*Dioscorea sambiranensis* R.Knuth in H.G.A.Engler (ed.), Pflanzenr., IV, 43: 353 (1924).
Ivory Coast, NW. Madagascar. 22 IVO 29 MDG. Cl. tuber geophyte.

*Dioscorea sanchez-colini* Matuda, Anales Inst. Biol. Univ. Nac. México 24: 336 (1954).
C. Mexico. 79 MXS. Cl. tuber geophyte.

*Dioscorea sandiensis* R.Knuth, Notizbl. Bot. Gart. Berlin-Dahlem 7: 192 (1917).
Peru. 83 PER. Cl. tuber geophyte.

*Dioscorea sandwithii* B.G.Schub., J. Arnold Arbor. 47: 158 (1966).
Belize. 80 BLZ. Cl. tuber geophyte.

*Dioscorea sanpaulensis* R.Knuth in H.G.A.Engler (ed.), Pflanzenr., IV, 43: 57 (1924).
SE. Brazil. 84 BZL. Cl. tuber geophyte.

*Dioscorea sansibarensis* Pax, Bot. Jahrb. Syst. 15: 146 (1892).
Trop. Africa, Madagascar. 22 BEN IVO NGA TOG 23 CAF CMN GAB ZAI 24 SUD 25 KEN TAN UGA 26 ANG MLW MOZ ZAM ZIM 29 MDG (78) fla. Cl. tuber geophyte.
*Dioscorea toxicaria* Bojer, Hortus Maurit.: 352 (1837), nom. nud.
*Dioscorea macroura* Harms, Notizbl. Königl. Bot. Gart. Berlin 2: 266 (1896).
*Dioscorea welwitschii* Rendle in W.P.Hiern, Cat. Afr. Pl. 2: 39 (1899).
*Dioscorea macabiha* Jum. & H.Perrier, Compt. Rend. Hebd. Séances Acad. Sci. 149: 485 (1909).
*Dioscorea maciba* Jum. & H.Perrier, Compt. Rend. Hebd. Séances Acad. Sci. 149: 486 (1909).

*Dioscorea santanderensis* R.Knuth, Repert. Spec. Nov. Regni Veg. 42: 161 (1937).
Colombia. 83 CLM. Cl. tuber geophyte.

*Dioscorea santosensis* R.Knuth, Repert. Spec. Nov. Regni Veg. 21: 77 (1925).
Brazil (São Paulo). 84 BZL. Cl. tuber geophyte.

*Dioscorea sarasinii* Uline ex R.Knuth in H.G.A.Engler (ed.), Pflanzenr., IV, 43: 291 (1924).
N. Sulawesi. 42 SUL. Cl. tuber geophyte.

*Dioscorea saxatilis* Poepp., Fragm. Syn. Pl.: 11 (1833).
C. Chile. 85 CLC. Cl. tuber geophyte.
*Dioscorea heterophylla* Poepp., Fragm. Syn. Pl.: 11 (1833), nom. illeg. *Dioscorea arenaria* Kunth, Enum. Pl. 5: 344 (1850).
*Dioscorea linearis* Bertero ex Colla, Mem. Reale Accad. Sci. Torino 39: 11 (1836).
*Dioscorea longifolia* Phil., Linnaea 33: 254 (1865), nom. illeg. *Dioscorea heterophylla* var. *longifolia* L.E. Navas, Anales Acad. Chilena Ci. Nat. 1968: 46 (1968).
*Dioscorea parviflora* Phil., Linnaea 33: 257 (1865).

*Dioscorea scabra* Humb. & Bonpl. ex Willd., Sp. Pl. 4: 794 (1806). *Helmia scabra* (Humb. & Bonpl. ex Willd.) Kunth, Enum. Pl. 5: 430 (1850).
S. Venezuela to Paraguay. 82 VEN 84 BZC BZE BZL BZN BZS 85 AGE? PAR. Cl. tuber geophyte.

*Dioscorea schimperiana* Hochst. ex Kunth, Enum. Pl. 5: 339 (1850).
Trop. Africa. 22 NGA 23 BUR CMN GAB RWA ZAI 24 ETH SUD 25 KEN TAN UGA 26 MLW MOZ ZIM. Cl. tuber geophyte.
*Dioscorea schimperiana* var. *vestita* Pax, Bot. Jahrb. Syst. 15: 148 (1892).
*Dioscorea fulvida* Stapf, J. Linn. Soc., Bot. 37: 530 (1906).
*Dioscorea hockii* De Wild., Bull. Jard. Bot. État 3: 277 (1911).
*Dioscorea stellatopilosa* De Wild., Ann. Mus. Congo Belge, Bot., V, 3: 369 (1912).
*Dioscorea stellatopilosa* var. *cordata* De Wild., Ann. Mus. Congo Belge, Bot., V, 3: 370 (1912).
*Dioscorea schimperiana* var. *adamaowense* Jacq.-Fél., Rev. Int. Bot. Appl. Agric. Trop. 27: 126 (1947).

*Dioscorea schubertiae* Ayala, Phytologia 55: 296 (1984).
Peru. 83 PER. Cl. tuber geophyte.
*\*Dioscorea elegans* R.Knuth, Repert. Spec. Nov. Regni Veg. 28: 83 (1930), nom. illeg.

*Dioscorea schunkei* Ayala & T.Clayton, Ann. Missouri Bot. Gard. 68: 130 (1981).
Peru. 83 PER. Cl. tuber geophyte.

*Dioscorea schwackei* Uline ex R.Knuth, Notizbl. Bot. Gart. Berlin-Dahlem 7: 195 (1917).
Brazil (Minas Gerais). 84 BZL. Cl. tuber geophyte.

*Dioscorea scortechinii* Prain & Burkill, J. Proc. Asiat. Soc. Bengal 4: 455 (1908).
Bangladesh to Sumatera. 36 CHH 40 BAN 41 THA VIE 42 MLY SUM. Cl. tuber geophyte.
*Dioscorea scortechinii* var. *parviflora* Prain & Burkill, Bull. Misc. Inform. Kew 1936: 494 (1936).

*Dioscorea secunda* R.Knuth in H.G.A.Engler (ed.), Pflanzenr., IV, 43: 350 (1924).
Brazil (São Paulo). 84 BZL. Cl. tuber geophyte.

*Dioscorea sellowiana* Uline ex R.Knuth, Notizbl. Bot. Gart. Berlin-Dahlem 7: 199 (1917).

SE. Brazil. 84 BZL. Cl. tuber geophyte.
*Dioscorea sellowiana* var. *mantiqueirensis* R.Knuth in H.G.A.Engler (ed.), Pflanzenr., IV, 43: 118 (1924).

**Dioscorea semperflorens** Uline in H.G.A.Engler & K.A.E. Prantl, Nat. Pflanzenfam., Nachtr. 1: 87 (1897).
WC. Trop. Africa. 23 CAF CMN GAB ZAI. Cl. tuber geophyte.
*Dioscorea hypotricha* Uline in H.G.A.Engler & K.A.E.Prantl, Nat. Pflanzenfam., Nachtr. 1: 87 (1897).
*Dioscorea schlechteri* Harms in F.R.R.Schechter, Westafr. Kautschuk-Exped.: 273 (1900), nom. inval.

**Dioscorea septemloba** Thunb. in J.A.Murray, Syst. Veg. ed. 14: 889 (1784).
China to C. & S. Japan. 36 CHS 38 JAP KOR. Cl. rhizome geophyte.
*Dioscorea sititoana* Honda & Jôtani, J. Jap. Bot. 14: 235 (1938).
*Dioscorea septemloba* var. *platyphylla* M.Mizush. ex T.Shimizu, Fl. Nagano Pref.: 1511 (1997).

**Dioscorea septemnervis** Vell., Fl. Flumin. 10: t. 119 (1831).
Brazil (Rio de Janeiro). 84 BZL. Cl. tuber geophyte.

**Dioscorea sericea** R.Knuth in H.G.A.Engler (ed.), Pflanzenr., IV, 43: 65 (1924).
Colombia. 83 CLM. Cl. tuber geophyte.

**Dioscorea seriflora** Jum. & H.Perrier, Ann. Inst. Bot.-Géol. Colon. Marseille, II, 8: 161 (1910).
E. Madagascar. 29 MDG. Tuber geophyte.
*Dioscorea ovifotsy* H.Perrier, Mém. Soc. Linn. Normandie, Bot. 1(1): 18 (1928).

**Dioscorea serpenticola** Hoque & P.K.Mukh., J. Bombay Nat. Hist. Soc. 99: 371 (2002).
Andaman Is., Nicobar Is. 41 AND NCB. Cl. tuber geophyte.

**Dioscorea sessiliflora** McVaugh, in Fl. Novo-Galiciana 15: 384 (1989).
SW. Mexico. 79 MXS. Cl. tuber geophyte.

**Dioscorea sexrimata** Burkill, Kew Bull. 5: 259 (1950).
SE. Sulawesi. 42 SUL. Cl. tuber geophyte.

**Dioscorea simulans** Prain & Burkill, Bull. Misc. Inform. Kew 1931: 427 (1931).
SE. China. 36 CHS. Cl. rhizome geophyte.

**Dioscorea sincorensis** R.Knuth, Notizbl. Bot. Gart. Berlin-Dahlem 7: 186 (1917).
WC. & E. Brazil. 84 BZC BZE BZL. Cl. tuber geophyte.

**Dioscorea sinoparviflora** C.T.Ting, M.G.Gilbert & Turland, Novon 10: 13 (2000).
China (Yunnan). 36 CHC. Cl. rhizome geophyte.
*Dioscorea parviflora* C.T.Ting, Acta Phytotax. Sin. 17(3): 69 (1979), nom. illeg.

**Dioscorea sinuata** Vell., Fl. Flumin. 10: t. 129 (1831).
Brazil to N. Argentina. 80 COS? 83 BOL 84 BZC BZE BZL BZS 85 AGE AGW CLN? PAR URU. Cl. tuber geophyte.
*Dioscorea crenata* Vell., Fl. Flumin. 10: t. 127 (1831).
*Dioscorea bonariensis* Ten., Index Seminum (NAP) 1838: 3 (1838).
*Dioscorea septemloba* Griseb. in C.F.P.von Martius & auct. suc. (eds.), Fl. Bras. 3(1): 46 (1842), nom. inval.

*Dioscorea variifolia* Kunze, Linnaea 20: 12 (1847), nom. illeg.
*Dioscorea sinuata* var. *bonariensis* Hauman, Anales Mus. Nac. Hist. Nat. Buenos Aires 27: 499 (1916).
*Dioscorea sinuata* var. *macrotepala* Uline, Notizbl. Bot. Gart. Berlin-Dahlem 7: 207 (1917).
*Dioscorea sinuata* var. *pauloensis* R.Knuth, Repert. Spec. Nov. Regni Veg. 36: 126 (1934).

**Dioscorea sitamiana** Prain & Burkill, Bull. Misc. Inform. Kew 1925: 64 (1925).
Borneo. 42 BOR. Cl. tuber geophyte.

**Dioscorea skottsbergii** R.Knuth, Repert. Spec. Nov. Regni Veg. 36: 125 (1934).
Chile. 85+. Cl. tuber geophyte.

**Dioscorea smilacifolia** De Wild. & T.Durand, Ann. Mus. Congo Belge, Bot., II, 1(1): 58 (1899).
W. Trop. Africa to Uganda and Angola. 22 BEN GHA IVO LBR NGA SIE TOG 23 CAF CMN GAB GGI ZAI 25 UGA 26 ANG. Cl. tuber geophyte.
*Dioscorea demeusei* De Wild. & T.Durand, Ann. Mus. Congo Belge, Bot., III, 1: 238 (1901).
*Dioscorea echinulata* De Wild., Bull. Jard. Bot. État 3: 278 (1911).
*Dioscorea flamignii* De Wild., Ann. Mus. Congo Belge, Bot., V, 3: 360 (1912).
*Dioscorea orbicularis* A.Chev. ex De Wild., Bull. Jard. Bot. État 4: 345 (1914), nom. inval.

**Dioscorea sonlaensis** R.Knuth, Repert. Spec. Nov. Regni Veg. 42: 162 (1937).
N. Vietnam. 41 VIE. Cl. tuber geophyte.

**Dioscorea sororopana** Steyerm., Fieldiana, Bot. 28(1): 159 (1951).
Venezuela (Bolívar). 82 VEN. Cl. tuber geophyte.

**Dioscorea soso** Jum. & H.Perrier, Compt. Rend. Hebd. Séances Acad. Sci. 149: 484 (1909).
Madagascar. 29 MDG. Tuber geophyte.

var. **soso**.
W. Madagascar. 29 MDG. Tuber geophyte.

var. **trichopoda** (Jum. & H.Perrier) Burkill & H.Perrier, in Fl. Madag. 44: 43 (1950).
Madagascar. 29 MDG. Cl. tuber geophyte.
*Dioscorea trichopoda* Jum. & H.Perrier, Ann. Inst. Bot.-Géol. Colon. Marseille, II, 8: 401 (1910).

**Dioscorea spectabilis** R.Knuth, Meded. Rijks-Herb. 29: 55 (1916).
Peru to Argentina (Jujuy). 83 BOL PER 85 AGW. Cl. tuber geophyte.

**Dioscorea spicata** Roth, Nov. Pl. Sp.: 371 (1821).
S. India, Sri Lanka, SE. Bangladesh. 40 BAN IND SRL. Cl. tuber geophyte.
*Dioscorea nummularia* var. *glauca* Prain & Burkill, J. Proc. Asiat. Soc. Bengal 10: 35 (1914).
*Dioscorea spicata* var. *anamallayana* Prain & Burkill, J. Proc. Asiat. Soc. Bengal 10: 29 (1914).
*Dioscorea spicata* var. *parvifolia* Prain & Burkill, J. Proc. Asiat. Soc. Bengal 10: 29 (1914).

**Dioscorea spiculiflora** Hemsl., Biol. Cent.-Amer., Bot. 3: 361 (1884).
Mexico to C. America. 79 MXE MXG MXS MXT 80 BLZ COS GUA HON NIC PAN. Cl. tuber geophyte.
*Dioscorea friedrichsthalii* R.Knuth in H.G.A.Engler (ed.), Pflanzenr., IV, 43: 169 (1924).

*Dioscorea spiculoides* Matuda, Anales Inst. Biol. Univ. Nac. México 24: 57 (1953).
C. Mexico. 79 MXC. Cl. tuber geophyte.

*Dioscorea spongiosa* J.Q.Xi, M.Mizuno & W.L.Zhao, Acta Phytotax. Sin. 25: 52 (1987).
SE. China (to SW. Hubei). 36 CHC CHS. Cl. rhizome geophyte.

*Dioscorea sprucei* Uline ex R.Knuth, Notizbl. Bot. Gart. Berlin-Dahlem 7: 197 (1917).
Ecuador to N. Peru. 83 ECU PER. Cl. tuber geophyte.

*Dioscorea standleyi* C.V.Morton, Publ. Carnegie Inst. Wash. 461: 252 (1936).
Costa Rica to Panama. 80 COS PAN. Cl. tuber geophyte.

*Dioscorea stegelmanniana* R.Knuth, Notizbl. Bot. Gart. Berlin-Dahlem 7: 203 (1917).
Brazil to Bolivia. 83 BOL PER 84 BZC BZN. Cl. tuber geophyte.
  *Dioscorea ferruginicaulis* Rusby, Mem. New York Bot. Gard. 7: 217 (1927).
  *Dioscorea ainensis* R.Knuth, Repert. Spec. Nov. Regni Veg. 29: 93 (1931).

*Dioscorea stellaris* R.Knuth in H.G.A.Engler (ed.), Pflanzenr., IV, 43: 233 (1924).
Brazil (Rio de Janeiro). 84 BZL. Cl. tuber geophyte.

*Dioscorea stemonoides* Prain & Burkill, Bull. Misc. Inform. Kew 1927: 244 (1927).
Thailand. 41 THA. Cl. tuber geophyte.

*Dioscorea stenocolpus* Phil., Linnaea 33: 258 (1865).
C. Chile. 85 CLC. Cl. tuber geophyte.

*Dioscorea stenomeriflora* Prain & Burkill, J. Proc. Asiat. Soc. Bengal 10: 40 (1914).
Pen. Malaysia to E. Sumatera. 42 BOR? MLY SUM. Cl. tuber geophyte.

*Dioscorea stenopetala* Hauman, Anales Mus. Nac. Hist. Nat. Buenos Aires 27: 505 (1916).
NW. Argentina. 85 AGW. Cl. tuber geophyte.
  *Dioscorea tucumanensis* R.Knuth, Notizbl. Bot. Gart. Berlin-Dahlem 7: 199 (1917).

*Dioscorea stenophylla* Uline in H.G.A.Engler & K.A.E.Prantl, Nat. Pflanzenfam., Nachtr. 1: 83 (1897).
Brazil (Minas Gerais to Goiás). 84 BZC BZL. Tuber geophyte.
  *Dioscorea stenophylla* var. *paucinervis* Uline ex R.Knuth, Notizbl. Bot. Gart. Berlin-Dahlem 7: 189 (1917).

*Dioscorea sterilis* O.Weber & Wilkin, Kew Bull. 60: 286 (2005).
Madagascar. 29 MDG.

*Dioscorea stipulosa* Uline ex R.Knuth in H.G.A.Engler (ed.), Pflanzenr., IV, 43: 94 (1924).
Cape Prov. 27 CPP. Cl. tuber geophyte.

*Dioscorea subcalva* Prain & Burkill, J. Proc. Asiat. Soc. Bengal 10: 18 (1914).
S. China. 36 CHC CHS. Cl. tuber geophyte.
  *Dioscorea submollis* R.Knuth in H.G.A.Engler (ed.), Pflanzenr., IV, 43: 318 (1924). *Dioscorea subcalva* var. *submollis* (R.Knuth) C.T.Ting & P.P.Ling, in Fl. Reipubl. Popul. Sin. 16(1): 85 (1985).

*Dioscorea subhastata* Vell., Fl. Flumin. 10: t. 121 (1831).
Brazil to NE. Argentina and Bolivia. 83 BOL 84 BZC

BZE BZL BZS 85 AGE PAR URU. Cl. tuber geophyte.
  *Dioscorea guaranitica* Chodat & Hassl., Bull. Herb. Boissier, II, 3: 1112 (1903).
  *Dioscorea guaranitica* f. *membranacea* Chodat & Hassl., Bull. Herb. Boissier, II, 3: 1112 (1903).
  *Dioscorea guaranitica* f. *subcoriacea* Chodat & Hassl., Bull. Herb. Boissier, II, 3: 1112 (1903).
  *Dioscorea lagoa-santa* Uline ex R.Knuth, Notizbl. Bot. Gart. Berlin-Dahlem 7: 201 (1917).
  *Dioscorea guaranitica* var. *balansae* Pellegr., Bull. Soc. Bot. Genève, II, 10: 387 (1919).
  *Dioscorea monadelpha* var. *opaca* Hicken, Darwiniana 1: 115 (1924).
  *Dioscorea piratinyensis* R.Knuth, Repert. Spec. Nov. Regni Veg. 22: 346 (1926).

*Dioscorea sublignosa* R.Knuth in H.G.A.Engler (ed.), Pflanzenr., IV, 43: 304 (1924).
Tanzania. 25 TAN. Cl. tuber geophyte. Provisionally accepted.

*Dioscorea submigra* R.Knuth, Repert. Spec. Nov. Regni Veg. 36: 126 (1934).
Dominican Rep. 81 DOM. Cl. tuber geophyte.

*Dioscorea subtomentosa* Miranda, Anales Inst. Biol. Univ. Nac. México 12: 583 (1941).
C. & S. Mexico. 79 MXC MXS MXT. Cl. tuber geophyte.
  *Dioscorea alboholosericea* Matuda, Bol. Soc. Bot. México 21: 2 (1957).

*Dioscorea sumatrana* Prain & Burkill, Bull. Misc. Inform. Kew 1931: 90 (1931).
E. Sumatera, Borneo. 42 BOR SUM. Cl. tuber geophyte.

*Dioscorea sumiderensis* B.G.Schub. & O.Téllez, Ann. Missouri Bot. Gard. 78: 248 (1991).
SE. Mexico. 79 MXT. Cl. tuber geophyte.

*Dioscorea suratensis* R.Knuth, Repert. Spec. Nov. Regni Veg. 28: 87 (1930).
Colombia. 83 CLM. Cl. tuber geophyte.

*Dioscorea sylvatica* Eckl., S. African Quart. J. 1: 363 (1830). *Testudinaria sylvatica* (Eckl.) Kunth, Enum. Pl. 5: 443 (1850).
S. Trop. & S. Africa. 26 MOZ ZAM ZIM 27 CPP NAT OFS SWZ TVL. Cl. tuber geophyte.
  *Dioscorea hederifolia* Griseb. in C.F.P.von Martius & auct. suc. (eds.), Fl. Bras. 3(1): 42 (1842).
  *Tamus sylvestris* Kunth, Enum. Pl. 5: 443 (1850), pro syn.
  *Dioscorea rehmannii* Baker in W.H.Harvey & auct. suc. (eds.), Fl. Cap. 6: 248 (1896). *Dioscorea sylvatica* var. *rehmannii* (Baker) Burkill, J. S. African Bot. 18: 189 (1952). *Testudinaria rehmannii* (Baker) G.D.Rowley, Natl. Cact. Succ. J. 8: 50 (1953). *Testudinaria sylvatica* var. *rehmannii* (Baker) G.D.Rowley, Natl. Cact. Succ. J. 28: 6 (1973).
  *Testudinaria montana* var. *paniculata* Kuntze, Revis. Gen. Pl. 3(2): 312 (1898). *Dioscorea montana* var. *paniculata* (Kuntze) R.Knuth in H.G.A.Engler (ed.), Pflanzenr., IV, 43: 323 (1924).
  *Testudinaria paniculata* Dummer, Bull. Misc. Inform. Kew 1912: 195 (1912). *Dioscorea montana* var. *duemmeri* R.Knuth in H.G.A.Engler (ed.), Pflanzenr., IV, 43: 323 (1924). *Dioscorea sylvatica* var. *paniculata* (Dummer) Burkill, J. S. African Bot. 18: 189 (1952). *Testudinaria sylvatica* var.

*paniculata* (Dummer) G.D.Rowley, Natl. Cact. Succ. J. 28: 6 (1973).
*Testudinaria multiflora* Marloth, Trans. Roy. Soc. South Africa 3: 127 (1913). *Dioscorea marlothii* R.Knuth in H.G.A.Engler (ed.), Pflanzenr., IV, 43: 321 (1924). *Dioscorea sylvatica* var. *multiflora* (Marloth) Burkill, J. S. African Bot. 18: 189 (1952). *Testudinaria sylvatica* var. *multiflora* (Marloth) G.D.Rowley, Natl. Cact. Succ. J. 28: 6 (1973). *Dioscorea brevipes* Burtt Davy, Bull. Misc. Inform. Kew 1924: 232 (1924). *Dioscorea sylvatica* var. *brevipes* (Burtt Davy) Burkill, J. S. African Bot. 18: 189 (1952). *Testudinaria sylvatica* var. *brevipes* (Burtt Davy) G.D.Rowley, Natl. Cact. Succ. J. 28: 6 (1973). *Dioscorea junodii* Burtt Davy, Bull. Misc. Inform. Kew 1924: 231 (1924). *Dioscorea montana* var. *glauca* R.Knuth in H.G.A.Engler (ed.), Pflanzenr., IV, 43: 323 (1924). *Dioscorea sylvatica* subsp. *lydenbergensis* Blunden, Hardman & F.J.Hind, Bot. J. Linn. Soc. 64: 445 (1971). *Testudinaria sylvatica* var. *lydenbergensis* (Blunden, Hardman & F.J.Hind) G.D.Rowley, Repert. Pl. Succ. 23: 11 (1972 publ. 1974).

**Dioscorea synandra** Uline in H.G.A.Engler & K.A.E.Prantl, Nat. Pflanzenfam., Nachtr. 1: 86 (1897). Brazil. 84 BZC BZL. Cl. tuber geophyte.

**Dioscorea syringifolia** (Kunth) Kunth & R.H.Schomb. ex R.Knuth in H.G.A.Engler (ed.), Pflanzenr., IV, 43: 107 (1924).
N. South America to Peru. 82 FRG GUY 83 PER 84 BZN. Cl. tuber geophyte.
*\*Helmia syringifolia* Kunth, Enum. Pl. 5: 423 (1850).

**Dioscorea tacanensis** Lundell, Lloydia 2: 78 (1939). Mexico (Chiapas). 79 MXT. Cl. tuber geophyte.

**Dioscorea tamarisciflora** Prain & Burkill, J. Proc. Asiat. Soc. Bengal 10: 22 (1914).
Pen. Thailand to Pen. Malaysia. 41 THA 42 MLY. Cl. tuber geophyte.

**Dioscorea tamoidea** Griseb. in C.F.P.von Martius & auct. suc. (eds.), Fl. Bras. 3(1): 42 (1842).
Cuba, Haiti. 81 CUB HAI. Cl. tuber geophyte.
*Dioscorea tamoidea* var. *lindenii* Uline, Notizbl. Bot. Gart. Berlin-Dahlem 7: 209 (1917).

**Dioscorea tamshiyacuensis** Ayala, Ann. Missouri Bot. Gard. 68: 125 (1981).
Peru. 83 PER. Cl. tuber geophyte.

**Dioscorea tanalarum** H.Perrier, Mém. Soc. Linn. Normandie, Bot. 1(1): 28 (1928).
Madagascar. 29 MDG. Cl. tuber geophyte.

**Dioscorea tancitarensis** Matuda, Anales Inst. Biol. Univ. Nac. México 24: 58 (1953).
SW. Mexico. 79 MXS. Cl. tuber geophyte.

**Dioscorea tarijensis** R.Knuth, Notizbl. Bot. Gart. Berlin-Dahlem 7: 212 (1917).
S. Bolivia. 83 BOL. Cl. tuber geophyte.

**Dioscorea tarmensis** R.Knuth, Notizbl. Bot. Gart. Berlin-Dahlem 7: 188 (1917).
Peru. 83 PER. Cl. tuber geophyte.

**Dioscorea tauriglossum** R.Knuth in H.G.A.Engler (ed.), Pflanzenr., IV, 43: 350 (1924).
Brazil (São Paulo). 84 BZL. Cl. tuber geophyte.

**Dioscorea tayacajensis** R.Knuth, Notizbl. Bot. Gart. Berlin-Dahlem 7: 212 (1917).
Peru. 83 PER. Cl. tuber geophyte.

**Dioscorea temascaltepecensis** R.Knuth, Repert. Spec. Nov. Regni Veg. 40: 223 (1936).
C. Mexico. 79 MXC. Cl. tuber geophyte.

**Dioscorea tenebrosa** C.V.Morton, Publ. Carnegie Inst. Wash. 461: 247 (1936).
Guatemala. 80 GUA. Cl. tuber geophyte.

**Dioscorea tenella** Phil., Fl. Atacam.: 51 (1860).
N. Chile. 85 CLN. Tuber geophyte.

**Dioscorea tentaculigera** Prain & Burkill, J. Proc. Asiat. Soc. Bengal 10: 15 (1914).
China (SW. Yunnan) to N. Thailand. 36 CHC 41 MYA THA. Cl. tuber geophyte.

**Dioscorea tenuifolia** Ridl., J. Straits Branch Roy. Asiat. Soc. 41: 34 (1904). *Dioscorea orbiculata* var. *tenuifolia* (Ridl.) Thapyai, Thai Forest Bull., Bot. 33: 197 (2005).
Pen. Thailand to Sumatera. 41 THA 42 MLY SUM. Cl. tuber geophyte.

**Dioscorea tenuipes** Franch. & Sav., Enum. Pl. Jap. 2: 523 (1878).
SE. China to C. & S. Japan. 36 CHS 38 JAP KOR. Cl. rhizome geophyte.
*Dioscorea acrotheca* Uline ex R.Knuth in H.G.A.Engler (ed.), Pflanzenr., IV, 43: 178 (1924), nom. inval.
*Dioscorea maximowiczii* Uline ex R.Knuth in H.G.A.Engler (ed.), Pflanzenr., IV, 43: 178 (1924).

**Dioscorea tenuiphyllum** R.Knuth in H.G.A.Engler (ed.), Pflanzenr., IV, 43: 76 (1924).
Brazil (?). 84+. Cl. tuber geophyte.

**Dioscorea tenuis** R.Knuth, Notizbl. Bot. Gart. Berlin-Dahlem 7: 206 (1917).
S. Chile. 85 CLS. Cl. tuber geophyte.

**Dioscorea tequendamensis** R.Knuth, Repert. Spec. Nov. Regni Veg. 30: 159 (1932).
Colombia. 83 CLM. Cl. tuber geophyte.

**Dioscorea ternata** Griseb., Vidensk. Meddel. Naturhist. Foren. Kjøbenhavn 1875: 158 (1875).
SE. Brazil. 84 BZL. Cl. tuber geophyte.

**Dioscorea therezopolensis** Uline ex R.Knuth, Notizbl. Bot. Gart. Berlin-Dahlem 7: 211 (1917).
Brazil (Rio de Janeiro). 84 BZL. Cl. tuber geophyte.
*Dioscorea therezopolensis* var. *latifolia* Uline ex R.Knuth, Notizbl. Bot. Gart. Berlin-Dahlem 7: 211 (1917).

**Dioscorea togoensis** R.Knuth in H.G.A.Engler (ed.), Pflanzenr., IV, 43: 299 (1924).
W. Trop. Africa. 22 BEN GAM GHA GUI IVO NGA SEN SIE TOG. Cl. tuber geophyte.
*Dioscorea caillei* A.Chev. ex De Wild., Bull. Jard. Bot. État 4: 332 (1914).

**Dioscorea tokoro** Makino ex Miyabe, Bot. Mag. (Tokyo) 3: 112 (1889).
China to Japan. 36 CHC CHS 38 JAP KOR. Cl. rhizome geophyte.
*Dioscorea buergeri* var. *enneaneura* Uline ex Diels, Bot. Jahrb. Syst. 29: 260 (1900). *Dioscorea enneaneura* (Uline ex Diels) Prain & Burkill, J. Asiat. Soc. Bengal, Pt. 2, Nat. Hist. 73(Suppl.): 11 (1904).
*Dioscorea buergeri* Uline ex R.Knuth, J. Asiat. Soc. Bengal, Pt. 2, Nat. Hist. 73(Suppl.): 11 (1904).

*Dioscorea yokusai* Prain & Burkill, J. Asiat. Soc. Bengal, Pt. 2, Nat. Hist. 73(Suppl.): 10 (1904).
*Dioscorea saidae* R.Knuth in H.G.A.Engler (ed.), Pflanzenr., IV, 43: 317 (1924).
*Dioscorea wichurae* Uline ex R.Knuth in H.G.A.Engler (ed.), Pflanzenr., IV, 43: 316 (1924).

**Dioscorea toldosensis** R.Knuth in H.G.A.Engler (ed.), Pflanzenr., IV, 43: 351 (1924).
Bolivia to NW. Argentina (Salta). 83 BOL 85 AGW. Cl. tuber geophyte.
*\*Dioscorea violacea* R.Knuth, Notizbl. Bot. Gart. Berlin-Dahlem 7: 200 (1917), nom. illeg.

**Dioscorea tomentosa** J.König ex Spreng., Pl. Min. Cogn. Pug. 2: 92 (1815). *Helmia tomentosa* (J.König ex Spreng.) Kunth, Enum. Pl. 5: 439 (1850).
India to Bangladesh, Sri Lanka. 40 BAN IND SRL. Cl. tuber geophyte.

**Dioscorea torticaulis** R.Knuth in H.G.A.Engler (ed.), Pflanzenr., IV, 43: 351 (1924).
Brazil (São Paulo). 84 BZL. Cl. tuber geophyte.

**Dioscorea trachyandra** Griseb., Vidensk. Meddel. Naturhist. Foren. Kjøbenhavn 1875: 155 (1875).
Guyana to Brazil. 82 GUY 84 BZC BZL BZS. Cl. tuber geophyte.

**Dioscorea trachycarpa** Kunth in F.W.H.von Humboldt, A.J.A.Bonpland & C.S.Kunth, Nov. Gen. Sp. 1: 274 (1816).
Guyana. 82 GUY. Cl. tuber geophyte.

**Dioscorea traillii** R.Knuth in H.G.A.Engler (ed.), Pflanzenr., IV, 43: 75 (1924).
N. Brazil. 84 BZN. Cl. tuber geophyte.

**Dioscorea transversa** R.Br., Prodr.: 295 (1810).
N. & E. Australia. 50 NSW NTA QLD WAU. Cl. tuber geophyte.
*Dioscorea punctata* R.Br., Prodr.: 294 (1810).

**Dioscorea trichantha** Baker, J. Linn. Soc., Bot. 20: 271 (1883).
C. Madagascar. 29 MDG. Cl. tuber geophyte.

**Dioscorea trichanthera** Gleason, Bull. Torrey Bot. Club 52: 182 (1925).
S. Venezuela, Guyana, Brazil. 82 GUY VEN 84 BZN. Cl. tuber geophyte.

**Dioscorea trifida** L.f., Suppl. Pl.: 427 (1782).
Trop. America. 80 COS GUA HON NIC PAN (81) jam LEE pue TRT WIN 82 FRG GUY SUR VEN 83 CLM ECU PER 84 BZC BZE BZL BZN. Cl. tuber geophyte.
*Dioscorea triloba* Lam., Encycl. 3: 234 (1789).
*Dioscorea brasiliensis* Willd., Sp. Pl. 4: 791 (1806).
*Dioscorea palmata* Juss. ex Pers., Syn. Pl. 2: 621 (1806).
*Dioscorea triloba* Willd., Sp. Pl. 4: 791 (1806), nom. illeg.
*Dioscorea quinquelobata* Vell., Fl. Flumin. 10: t. 128 (1831), nom. illeg. *Dioscorea articulata* Steud., Nomencl. Bot., ed. 2, 1: 511 (1840).
*Dioscorea goyazensis* Griseb. in C.F.P.von Martius & auct. suc. (eds.), Fl. Bras. 3(1): 41 (1842).
*Dioscorea affinis* Kunth, Enum. Pl. 5: 372 (1850).
*Dioscorea ruiziana* Klotzsch ex Kunth, Enum. Pl. 5: 374 (1850).
*Dioscorea angustifolia* Rusby, Bull. Torrey Bot. Club 29: 701 (1902), nom. illeg.

**Dioscorea trifoliata** Kunth in F.W.H.von Humboldt, A.J.A.Bonpland & C.S.Kunth, Nov. Gen. Sp. 1: 275 (1816). *Helmia trifoliata* (Kunth) Kunth, Enum. Pl. 5: 417 (1850).
Trinidad, Panama to Peru. 80 PAN 81 TRT 82 VEN 83 CLM PER 84 BZN. Cl. tuber geophyte.
*Dioscorea galipanensis* Klotzsch ex Kunth, Enum. Pl. 5: 417 (1850), pro syn.
*Dioscorea triloba* H.Karst. ex Kunth, Enum. Pl. 5: 418 (1850), nom. inval.
*Helmia galipanensis* Kunth, Enum. Pl. 5: 417 (1850).
*Dioscorea trifoliata* var. *galipanensis* (Kunth) Uline ex R.Knuth, Notizbl. Bot. Gart. Berlin-Dahlem 7: 202 (1917).
*Dioscorea trifoliata* var. *amazonica* R.Knuth in H.G.A.Engler (ed.), Pflanzenr., IV, 43: 129 (1924).

**Dioscorea trifurcata** Hauman, Anales Mus. Nac. Hist. Nat. Buenos Aires 27: 482 (1916).
Argentina (Catamarca). 85 AGW. Cl. tuber geophyte.

**Dioscorea trilinguis** Griseb., Vidensk. Meddel. Naturhist. Foren. Kjøbenhavn 1875: 163 (1875).
SE. Brazil. 84 BZL. Cl. tuber geophyte.
*Dioscorea trilinguis* var. *edwallii* Uline ex R.Knuth, Notizbl. Bot. Gart. Berlin-Dahlem 7: 218 (1917).

**Dioscorea trimenii** Prain & Burkill, J. Proc. Asiat. Soc. Bengal 10: 29 (1914).
Sri Lanka. 40 SRL. Cl. tuber geophyte.

**Dioscorea trinervia** Roxb. ex Prain & Burkill, J. Proc. Asiat. Soc. Bengal 10: 32 (1914).
Assam to NW. Myanmar. 40 ASS BAN 41 MYA. Cl. tuber geophyte.

**Dioscorea trisecta** Griseb., Vidensk. Meddel. Naturhist. Foren. Kjøbenhavn 1875: 159 (1875).
Peru to Brazil. 83 PER 84 BZC BZL. Cl. tuber geophyte.
*Helmia grisebachii* Kunth, Enum. Pl. 5: 432 (1850).

**Dioscorea trollii** R.Knuth, Repert. Spec. Nov. Regni Veg. 30: 161 (1932).
Bolivia. 83 BOL. Cl. tuber geophyte.

**Dioscorea truncata** Miq., Linnaea 18: 23 (1844).
Suriname. 82 SUR. Cl. tuber geophyte. Provisionally accepted.

**Dioscorea tsaratananensis** H.Perrier, Mém. Soc. Linn. Normandie, Bot. 1(1): 20 (1928).
N. & NE. Madagascar. 29 MDG. Cl. tuber geophyte.
*Dioscorea knuthii* H.Perrier, Mém. Soc. Linn. Normandie, Bot. 1(1): 32 (1928).

**Dioscorea tubiperianthia** Matuda, Anales Inst. Biol. Univ. Nac. México 24: 55 (1953).
C. Mexico. 79 MXC. Cl. tuber geophyte.

**Dioscorea tubuliflora** Uline ex R.Knuth, Notizbl. Bot. Gart. Berlin-Dahlem 7: 189 (1917).
SE. Brazil. 84 BZL. Cl. tuber geophyte.

**Dioscorea tubulosa** Griseb., Vidensk. Meddel. Naturhist. Foren. Kjøbenhavn 1875: 154 (1875).
SE. Brazil. 84 BZL. Cl. tuber geophyte.

**Dioscorea uliginosa** Phil., Anales Univ. Chile 93: 12 (1896).
SC. Chile. 85 CLC. Tuber geophyte.

**Dioscorea ulinei** Greenm. ex R.Knuth, Notizbl. Bot. Gart. Berlin-Dahlem 7: 191 (1917).
C. Mexico. 79 MXC. Cl. tuber geophyte.

*Dioscorea ulinei* var. *longipes* Matuda, Anales Inst. Biol. Univ. Nac. México 24: 359 (1954).

**Dioscorea undatiloba** Baker, J. Bot. 27: 8 (1889).
Northern Prov. to KwaZulu-Natal. 27 NAT TVL. Cl. tuber geophyte.

**Dioscorea urceolata** Uline, Bot. Jahrb. Syst. 22: 426 (1896).
C. & SW. Mexico. 79 MXC MXS. Cl. tuber geophyte.
*Dioscorea urceolata* var. *reflexa* Greenm. ex R.Knuth in H.G.A.Engler (ed.), Pflanzenr., IV, 43: 229 (1924).
*Dioscorea urceolata* f. *atropurpureoloba* Matuda, Anales Inst. Biol. Univ. Nac. México 24: 329 (1954).

**Dioscorea urophylla** Hemsl., Biol. Cent.-Amer., Bot. 3: 361 (1884).
S. Mexico to Venezuela. 79 MXS 80 COS HON NIC PAN 82 VEN 83 CLM. Cl. tuber geophyte.

**Dioscorea uruapanensis** Matuda, Bol. Soc. Bot. México 15: 26 (1953).
C. & SW. Mexico. 79 MXC MXS. Cl. tuber geophyte.

**Dioscorea valdiviensis** R.Knuth, Notizbl. Bot. Gart. Berlin-Dahlem 7: 206 (1917).
S. Chile. 85 CLS. Cl. tuber geophyte.

**Dioscorea vanvuurenii** Prain & Burkill, Bull. Misc. Inform. Kew 1925: 63 (1925).
C. & S. Sulawesi. 42 SUL. Cl. tuber geophyte.

**Dioscorea variifolia** Bertero, Mercurio Chileno 13: 612 (1829).
C. Chile. 85 CLC. Cl. tuber geophyte.
*Dioscorea acuminata* Steud., Nomencl. Bot., ed. 2, 1: 510 (1840), nom. inval.
*Dioscorea microphylla* Steud., Nomencl. Bot., ed. 2, 1: 511 (1840), nom. inval.

**Dioscorea velutipes** Prain & Burkill, J. Proc. Asiat. Soc. Bengal 10: 19 (1914).
SC. China to N. Thailand. 36 CHC 41 MYA THA. Cl. tuber geophyte.

**Dioscorea vexans** Prain & Burkill, J. Proc. Asiat. Soc. Bengal 4: 456 (1908).
S. Andaman Is. 41 AND. Cl. tuber geophyte.

**Dioscorea vilis** Kunth, Enum. Pl. 5: 400 (1850).
Jawa. 42 JAW. Cl. tuber geophyte.

**Dioscorea villosa** L., Sp. Pl.: 1033 (1753). *Merione villosa* (L.) Salisb., Gen. Pl.: 12 (1866).
S. Ontario to C. & E. U.S.A. 72 ONT 74 ILL IOW KAN MIN MSO NEB OKL WIS 75 CNT INI MAS MIC NWJ NWY OHI PEN RHO VER WVA 77 TEX 78 ALA ARK DEL FLA GEO KTY LOU MRY MSI NCA SCA TEN VRG WDC 79 MXS?. Cl. rhizome geophyte.
*Dioscorea sativa* L., Sp. Pl.: 1033 (1753).
*Dioscorea quaternata* Walter, Fl. Carol.: 246 (1788).
*Dioscorea villosa* subsp. *quaternata* (Walter) R.Knuth in H.G.A.Engler (ed.), Pflanzenr., IV, 43: 173 (1924).
*Dioscorea quinata* Walter, Fl. Carol.: 246 (1788).
*Dioscorea cliffortiana* Lam., Encycl. 3: 232 (1789).
*Dioscorea paniculata* Michx., Fl. Bor.-Amer. 2: 239 (1803). *Dioscorea villosa* subsp. *paniculata* (Michx.) R.Knuth in H.G.A.Engler (ed.), Pflanzenr., IV, 43: 173 (1924).
*Dioscorea waltheri* Desf., Tabl. École Bot., ed. 2: 269 (1815).

*Dioscorea glauca* Muhl. ex L.C.Beck, Bot. North. Middle States: 355 (1833). *Dioscorea villosa* subsp. *glauca* (Muhl. ex L.C.Beck) R.Knuth in H.G.A.Engler (ed.), Pflanzenr., IV, 43: 173 (1924). *Dioscorea quaternata* var. *glauca* (Muhl. ex L.C.Beck) Fernald, Rhodora 39: 399 (1937).
*Dioscorea hexaphylla* Raf., New Fl. 2: 89 (1837).
*Dioscorea longifolia* Raf., New Fl. 2: 89 (1837).
*Dioscorea megaptera* Raf., New Fl. 2: 88 (1837).
*Dioscorea repanda* Raf., New Fl. 2: 89 (1837), nom. illeg.
*Dioscorea villosa* var. *laeviuscula* Alph.Wood, Classbook Bot., ed. 2: 544 (1847).
*Dioscorea pruinosa* Kunth, Enum. Pl. 5: 338 (1850).
*Dioscorea villosa* var. *glabra* J.Lloyd ex A.Gray, Manual, ed. 7: 297 (1908). *Dioscorea lloydiana* E.H.L.Krause, Beih. Bot. Centralbl. 32(2): 331 (1914).
*Dioscorea hirticaulis* Bartlett, Bull. Bur. Pl. Industr. U.S.D.A. 189: 17 (1910). *Dioscorea villosa* subsp. *hirticaulis* (Bartlett) R.Knuth in H.G.A.Engler (ed.), Pflanzenr., IV, 43: 173 (1924). *Dioscorea villosa* var. *hirticaulis* (Bartlett) H.E.Ahles, J. Elisha Mitchell Sci. Soc. 80: 172 (1964).
*Dioscorea paniculata* var. *glabrifolia* Bartlett, Bull. Bur. Pl. Industr. U.S.D.A. 189: 15 (1910). *Dioscorea villosa* subsp. *glabrifolia* (Bartlett) W.Stone, Pl. S. New Jersey: 358 (1912). *Dioscorea villosa* var. *glabrifolia* (Bartlett) S.F.Blake, Rhodora 20: 49 (1918). *Dioscorea villosa* f. *glabrifolia* (Bartlett) Fernald, Rhodora 39: 401 (1937).

**Dioscorea volckmannii** Phil., Linnaea 33: 255 (1865).
C. Chile. 85 CLC. Tuber geophyte.

**Dioscorea wallichii** Hook.f., Fl. Brit. India 6: 295 (1892).
India to China (W. Yunnan) and Pen. Malaysia. 36 CHC 40 ASS BAN EHM IND 41 MYA THA 42 MLY. Cl. tuber geophyte.
*Dioscorea wallichii* var. *christiei* Prain & Burkill, J. Proc. Asiat. Soc. Bengal 10: 31 (1914).
*Dioscorea wallichii* var. *vera* Prain & Burkill, J. Proc. Asiat. Soc. Bengal 10: 31 (1914).

**Dioscorea warburgiana** Uline ex Prain & Burkill, J. Proc. Asiat. Soc. Bengal 4: 456 (1908).
N. Sulawesi. 42 SUL. Cl. tuber geophyte.
*Dioscorea celebesiana* R.Knuth in H.G.A.Engler (ed.), Pflanzenr., IV, 43: 291 (1924).

**Dioscorea warmingii** R.Knuth in H.G.A.Engler (ed.), Pflanzenr., IV, 43: 355 (1924).
Brazil (Minas Gerais). 84 BZL. Cl. tuber geophyte.
*\*Dioscorea tenuifolia* Uline ex R.Knuth, Notizbl. Bot. Gart. Berlin-Dahlem 7: 204 (1917), nom. illeg.

**Dioscorea wattii** Prain & Burkill, J. Proc. Asiat. Soc. Bengal 4: 457 (1908).
E. Himalaya to Bangladesh. 40 ASS BAN EHM. Cl. tuber geophyte.

**Dioscorea weberbaueri** R.Knuth, Notizbl. Bot. Gart. Berlin-Dahlem 7: 205 (1917).
Peru. 83 PER. Cl. tuber geophyte.

**Dioscorea widgrenii** R.Knuth, Repert. Spec. Nov. Regni Veg. 21: 79 (1925).
Brazil (Rio de Janeiro). 84 BZL. Cl. tuber geophyte.

**Dioscorea wightii** Hook.f., Fl. Brit. India 6: 291 (1892).
S. India. 40 IND. Cl. tuber geophyte.

*Dioscorea wittiana* R.Knuth, Notizbl. Bot. Gart. Berlin-Dahlem 7: 194 (1917).
N. Brazil. 84 BZN. Cl. tuber geophyte.

*Dioscorea wrightii* Uline ex R.Knuth, Notizbl. Bot. Gart. Berlin-Dahlem 7: 208 (1917).
Cuba. 81 CUB. Cl. tuber geophyte.
   *Rajania herradurensis* R.Knuth, Notizbl. Bot. Gart. Berlin-Dahlem 7: 220 (1917). *Dioscorea herradurensis* (R.Knuth) P.Wilson ex Alain, Fl. Cuba 1: 321 (1946).

*Dioscorea xizangensis* C.T.Ting, Acta Phytotax. Sin. 20: 205 (1982).
Tibet. 36 CHT. Cl. tuber geophyte.

*Dioscorea yunnanensis* Prain & Burkill, J. Asiat. Soc. Bengal, Pt. 2, Nat. Hist. 73: 186 (1904).
SC. China. 36 CHC. Cl. tuber geophyte.

*Dioscorea zingiberensis* C.H.Wright, J. Linn. Soc., Bot. 36: 93 (1903).
C. China to Vietnam. 36 CHC CHN CHS 41 VIE. Cl. rhizome geophyte.
   *Dioscorea henryi* Uline ex Diels, Bot. Jahrb. Syst. 29: 261 (1900), nom. inval.

**Synonyms:**
*Dioscorea acarophyta* De Wild. = **Dioscorea minutiflora** Engl.
*Dioscorea acerifolia* Uline ex Prain & Burkill = **Dioscorea nipponica** Makino
*Dioscorea acerifolia* var. *rosthornii* Diels = **Dioscorea nipponica** Makino
*Dioscorea acrotheca* Uline ex R.Knuth = **Dioscorea tenuipes** Franch. & Sav.
*Dioscorea aculeata* Balb. ex Kunth = **Dioscorea cayennensis** Lam. subsp. *cayennensis*
*Dioscorea aculeata* Roxb. = **Dioscorea esculenta** (Lour.) Burkill
*Dioscorea aculeata* L. = [40 IND] Perhaps identical with Dioscorea esculenta.
*Dioscorea aculeata* var. *spinosa* Roxb. ex Prain & Burkill = **Dioscorea esculenta** (Lour.) Burkill
*Dioscorea acuminata* Steud. = **Dioscorea variifolia** Bertero
*Dioscorea acutata* R.Knuth = **Dioscorea coronata** Hauman
*Dioscorea acutifolia* Phil. = **Dioscorea auriculata** Poepp.
*Dioscorea adenocarpa* Mart. ex Griseb. = **Dioscorea ovata** Vell.
*Dioscorea adenocarpa* var. *balansae* Uline ex R.Knuth = **Dioscorea ovata** Vell.
*Dioscorea adenocarpa* var. *chartacea* Uline ex R.Knuth = **Dioscorea ovata** Vell.
*Dioscorea affinis* Kunth = **Dioscorea trifida** L.f.
*Dioscorea ainensis* R.Knuth = **Dioscorea stegelmanniana** R.Knuth
*Dioscorea alata* var. *globosa* (Roxb.) Prain = **Dioscorea alata** L.
*Dioscorea alata* var. *tarri* Prain & Burkill = **Dioscorea alata** L.
*Dioscorea alata* var. *vera* Prain & Burkill = **Dioscorea alata** L.
*Dioscorea albicaulis* Uline = **Dioscorea dicranandra** Donn.Sm.
*Dioscorea albinervia* R.Knuth = **Dioscorea marginata** Griseb.
*Dioscorea alboholosericea* Matuda = **Dioscorea subtomentosa** Miranda
*Dioscorea altissima* Sieber ex C.Presl = **Dioscorea polygonoides** Humb. & Bonpl. ex Willd.

*Dioscorea amaranthoides* var. *crumenigera* (Mart. ex Griseb.) Uline ex R.Knuth = **Dioscorea amaranthoides** C.Presl
*Dioscorea amaranthoides* var. *decorticans* (C.Presl) Uline ex R.Knuth = **Dioscorea decorticans** C.Presl
*Dioscorea amaranthoides* var. *denudata* Uline ex R.Knuth = **Dioscorea amaranthoides** C.Presl
*Dioscorea amaranthoides* var. *elegantula* Uline ex R.Knuth = **Dioscorea amaranthoides** C.Presl
*Dioscorea amaranthoides* var. *glauca* Uline ex R.Knuth = **Dioscorea amaranthoides** C.Presl
*Dioscorea amaranthoides* var. *metallica* Harms ex R.Knuth = **Dioscorea amaranthoides** C.Presl
*Dioscorea amaranthoides* var. *paniculata* R.Knuth = **Dioscorea amaranthoides** C.Presl
*Dioscorea amaranthoides* var. *pendula* (Poepp. ex Kunth) Uline ex R.Knuth = **Dioscorea pendula** Poepp. ex Kunth
*Dioscorea amaranthoides* var. *ulei* R.Knuth = **Dioscorea pendula** Poepp. ex Kunth
*Dioscorea amazonum* var. *burchellii* Uline ex R.Knuth = **Dioscorea amazonum** Mart. ex Griseb. var. *amazonum*
*Dioscorea amazonum* var. *consanguinea* (Kunth) Uline ex R.Knuth = **Dioscorea amazonum** Mart. ex Griseb. var. *amazonum*
*Dioscorea amazonum* var. *robustior* Uline ex R.Knuth = **Dioscorea amazonum** Mart. ex Griseb. var. *amazonum*
*Dioscorea amazonum* var. *sagotiana* Uline ex R.Knuth = **Dioscorea amazonum** Mart. ex Griseb. var. *amazonum*
*Dioscorea amazonum* var. *sprucei* Uline ex R.Knuth = **Dioscorea amazonum** Mart. ex Griseb. var. *amazonum*
*Dioscorea anchiatasi* Harms = **Dioscorea quartiniana** A.Rich.
*Dioscorea anchietae* Harms ex R.Knuth = **Dioscorea quartiniana** A.Rich.
*Dioscorea ancistrocarpa* Uline ex R.Knuth = **Dioscorea mandonii** Rusby
*Dioscorea anconensis* R.Knuth = **Dioscorea mexicana** Scheidw.
*Dioscorea andongensis* Rendle = **Dioscorea preussii** Pax subsp. *preussii*
*Dioscorea angolensis* R.Knuth = **Dioscorea quartiniana** A.Rich.
*Dioscorea anguina* Roxb. = **Dioscorea pubera** Blume
*Dioscorea angulata* R.Knuth = **Dioscorea nummularia** Lam.
*Dioscorea angusta* R.Knuth = **Dioscorea cirrhosa** Lour. var. *cirrhosa*
*Dioscorea angustiflora* Rendle = **Dioscorea praehensilis** Benth.
*Dioscorea angustifolia* Rusby = **Dioscorea trifida** L.f.
*Dioscorea angustifolia* Lam. = [83 PER]
*Dioscorea anthropophagorum* A.Chev. = **Dioscorea bulbifera** L.
*Dioscorea anthropophagorum* var. *sylvestris* A.Chev. = **Dioscorea hirtiflora** Benth. subsp. *hirtiflora*
*Dioscorea apaensis* Chodat & Hassl. = **Dioscorea amaranthoides** C.Presl
*Dioscorea apiculata* De Wild. = **Dioscorea quartiniana** A.Rich.
*Dioscorea apurimacensis* R.Knuth = **Dioscorea acanthogene** Rusby
*Dioscorea arcuata* Rusby = **Dioscorea mandonii** Rusby
*Dioscorea arenaria* Kunth = **Dioscorea saxatilis** Poepp.
*Dioscorea aristolochiifolia* var. *parvifolia* (Phil.) L.E.Navas = **Dioscorea modesta** Phil.

*Dioscorea armata* De Wild. = **Dioscorea minutiflora** Engl.

*Dioscorea articulata* Steud. = **Dioscorea trifida** L.f.

*Dioscorea astrostigma* Uline = **Dioscorea mexicana** Scheidw.

*Dioscorea atropurpurea* Roxb. = **Dioscorea alata** L.

*Dioscorea axilliflora* Phil. = **Dioscorea fastigiata** Gay

*Dioscorea balsapuertensis* R.Knuth = **Dioscorea altissima** Lam.

*Dioscorea bangii* R.Knuth = **Dioscorea dodecaneura** Vell.

*Dioscorea bararum* H.Perrier = **Dioscorea acuminata** Baker

*Dioscorea barclayi* R.Knuth = **Dioscorea floribunda** M.Martens & Galeotti

*Dioscorea batatas* Decne. = **Dioscorea polystachya** Turcz.

*Dioscorea batatas* f. *clavata* Makino = **Dioscorea polystachya** Turcz.

*Dioscorea batatas* f. *daikok* Makino = **Dioscorea polystachya** Turcz.

*Dioscorea batatas* f. *flabellata* Makino = **Dioscorea polystachya** Turcz.

*Dioscorea batatas* f. *rakuda* Makino = **Dioscorea polystachya** Turcz.

*Dioscorea batatas* f. *tsukune* Makino = **Dioscorea polystachya** Turcz.

*Dioscorea baya* var. *subcordata* De Wild. = **Dioscorea baya** De Wild. var. **baya**

*Dioscorea beccariana* Martelli = **Dioscorea quartiniana** A.Rich.

*Dioscorea belizensis* Lundell = **Dioscorea hondurensis** R.Knuth

*Dioscorea belophylloides* Prain & Burkill = **Dioscorea japonica** Thunb. var. **japonica**

*Dioscorea berteroana* Kunth = **Dioscorea cayennensis** Lam. subsp. **cayennensis**

*Dioscorea besseriana* var. *berteroi* Uline ex R.Knuth = **Dioscorea besseriana** Kunth

*Dioscorea bilbergiana* Kunth = **Dioscorea mexicana** Scheidw.

*Dioscorea biserialis* Prain & Burkill = **Dioscorea panthaica** Prain & Burkill

*Dioscorea boliviensis* R.Knuth = **Dioscorea glomerulata** Hauman

*Dioscorea bolojonica* Blanco = (Fabaceae)

*Dioscorea bonariensis* Ten. = **Dioscorea sinuata** Vell.

*Dioscorea bonatiana* Prain & Burkill = **Dioscorea kamoonensis** Kunth

*Dioscorea bonnetii* A.Chev. = **Dioscorea cirrhosa** Lour. var. **cirrhosa**

*Dioscorea borealis* C.V.Morton = **Dioscorea racemosa** (Klotzsch) Uline

*Dioscorea boridiensis* R.Knuth = [43 NWG]

*Dioscorea borneensis* R.Knuth = **Dioscorea piscatorum** Prain & Burkill

*Dioscorea brachybotrya* var. *germainii* Uline ex R.Knuth = **Dioscorea brachybotrya** Poepp.

*Dioscorea brachybotrya* var. *reticulata* (Gay) Uline ex R.Knuth = **Dioscorea reticulata** Gay

*Dioscorea brachycarpa* Schltdl. = **Dioscorea convolvulacea** Cham. & Schltdl. subsp. **convolvulacea**

*Dioscorea brasiliensis* Willd. = **Dioscorea trifida** L.f.

*Dioscorea brevifolia* (Prain & Burkill) K.Y.Guan & D.F.Chamb. = **Dioscorea kamoonensis** Kunth

*Dioscorea brevipes* Burtt Davy = **Dioscorea sylvatica** Eckl.

*Dioscorea brevispicata* De Wild. = **Dioscorea minutiflora** Engl.

*Dioscorea buchananii* var. *ukamensis* R.Knuth = **Dioscorea buchananii** Benth.

*Dioscorea buchholziana* Engl. = **Dioscorea dumetorum** (Kunth) Pax

*Dioscorea buchtienii* R.Knuth = **Dioscorea brachybotrya** Poepp.

*Dioscorea buergeri* Uline ex R.Knuth = **Dioscorea tokoro** Makino ex Miyabe

*Dioscorea buergeri* var. *enneaneura* Uline ex Diels = **Dioscorea tokoro** Makino ex Miyabe

*Dioscorea bulbifera* var. *anthropophagorum* (A.Chev.) Summerh. = **Dioscorea bulbifera** L.

*Dioscorea bulbifera* var. *crispata* (Roxb.) Prain = **Dioscorea bulbifera** L.

*Dioscorea bulbifera* var. *elongata* (F.M.Bailey) Prain & Burkill = **Dioscorea bulbifera** L.

*Dioscorea bulbifera* var. *pulchella* (Roxb.) Prain = **Dioscorea bulbifera** L.

*Dioscorea bulbifera* var. *sativa* Prain = **Dioscorea bulbifera** L.

*Dioscorea bulbifera* var. *suavia* Prain & Burkill = **Dioscorea bulbifera** L.

*Dioscorea bulbifera* var. *vera* Prain & Burkill = **Dioscorea bulbifera** L.

*Dioscorea bullata* Prain & Burkill = **Dioscorea flabellifolia** Prain & Burkill

*Dioscorea burkillii* R.Knuth = **Dioscorea delavayi** Franch.

*Dioscorea burroyacensis* R.Knuth = **Dioscorea glomerulata** Hauman

*Dioscorea caillei* A.Chev. ex De Wild. = **Dioscorea togoensis** R.Knuth

*Dioscorea calcarea* R.Knuth = **Dioscorea altissima** Lam.

*Dioscorea calcensis* R.Knuth = **Dioscorea glandulosa** var. *calcensis* (R.Knuth) Ayala

*Dioscorea calyculata* J.D.Sm. = (Basellaceae)

*Dioscorea calystegioides* Kunth = **Dioscorea laxiflora** Mart. ex Griseb.

*Dioscorea camerunensis* R.Knuth = **Dioscorea cayennensis** Lam. subsp. **cayennensis**

*Dioscorea campanulata* var. *lanceolata* Uline = **Dioscorea campanulata** Uline ex R.Knuth

*Dioscorea campestris* var. *grandiflora* Griseb. = **Dioscorea campestris** Griseb.

*Dioscorea campestris* f. *longispicata* Hauman = **Dioscorea campestris** Griseb.

*Dioscorea campestris* var. *longispicata* Hauman = **Dioscorea campestris** Griseb.

*Dioscorea campestris* f. *paraguayensis* R.Knuth = **Dioscorea campestris** Griseb.

*Dioscorea campestris* var. *parviflora* Griseb. = **Dioscorea campestris** Griseb.

*Dioscorea campestris* f. *pedalis* Uline ex R.Knuth = **Dioscorea campestris** Griseb.

*Dioscorea campestris* f. *piedadensis* Uline ex R.Knuth = **Dioscorea campestris** Griseb.

*Dioscorea campestris* f. *plantaginifolia* Uline ex R.Knuth = **Dioscorea campestris** Griseb.

*Dioscorea campestris* f. *stenorachis* Uline ex R.Knuth = **Dioscorea campestris** Griseb.

*Dioscorea camphorifolia* Uline ex R.Knuth = **Dioscorea cirrhosa** Lour. var. **cirrhosa**

*Dioscorea canariensis* Webb & Berthel. = **Dioscorea communis** (L.) Caddick & Wilkin

*Dioscorea capillaris* Hemsl. = **Dioscorea convolvulacea** Cham. & Schltdl. subsp. **convolvulacea**

*Dioscorea capillaris* var. *glabra* Hemsl. = **Dioscorea convolvulacea** Cham. & Schltdl. subsp. **convolvulacea**

*Dioscorea caracasana* Kunth = **Dioscorea polygonoides** Humb. & Bonpl. ex Willd.

*Dioscorea caracasensis* R.Knuth = *Dioscorea birschelii* Harms ex R.Knuth

*Dioscorea carrii* R.Knuth = *Dioscorea opaca* R.Knuth

*Dioscorea catharinensis* R.Knuth = *Dioscorea catharinensis* R.Knuth

*Dioscorea caucensis* R.Knuth = *Dioscorea coriacea* Humb. & Bonpl. ex Willd.

*Dioscorea cayennensis* var. *praehensilis* (Benth.) A.Chev. = *Dioscorea praehensilis* Benth.

*Dioscorea cayennensis* var. *pseudobatatas* Hauman = *Dioscorea polystachya* Turcz.

*Dioscorea celebesiana* R.Knuth = *Dioscorea warburgiana* Uline ex Prain & Burkill

*Dioscorea chamela* McVaugh = *Dioscorea convolvulacea* Cham. & Schltdl. subsp. *convolvulacea*

*Dioscorea changjiangensis* F.W.Xing & Z.X.Li = *Dioscorea pentaphylla* L.

*Dioscorea chevalieri* De Wild. = *Dioscorea preussii* Pax subsp. *preussii*

*Dioscorea chiquiacensis* R.Knuth = *Dioscorea hieronymi* Uline ex R.Knuth

*Dioscorea chondrocarpa* Griseb. = *Dioscorea altissima* Lam.

*Dioscorea cinnamomifolia* var. *zanoniae* (Klotzsch ex Griseb.) R.Knuth = *Dioscorea cinnamomifolia* Hook.

*Dioscorea clarkei* Prain & Burkill = *Dioscorea prazeri* Prain & Burkill

*Dioscorea clemensii* R.Knuth = [42 BOR]

*Dioscorea cliffortiana* Lam. = *Dioscorea villosa* L.

*Dioscorea codonopsidifolia* Kamik. = *Dioscorea pentaphylla* L.

*Dioscorea collinsae* Prain & Burkill = *Dioscorea arachidna* Prain & Burkill

*Dioscorea colocasiifolia* Pax = *Dioscorea alata* L.

*Dioscorea combilium* Buch.-Ham. = *Dioscorea pubera* Blume

*Dioscorea concepcionis* Chodat & Hassl. = *Dioscorea multiflora* Mart. ex Griseb.

*Dioscorea conferta* Vell. = *Dioscorea piperifolia* Humb. & Bonpl. ex Willd.

*Dioscorea congestiflora* R.Knuth = *Dioscorea ceratandra* Uline ex R.Knuth

*Dioscorea convolvulacea* subsp. *esurientium* (Uline) Uline ex R.Knuth = *Dioscorea convolvulacea* Cham. & Schltdl. subsp. *convolvulacea*

*Dioscorea convolvulacea* var. *galeottiana* (Kunth) Uline = *Dioscorea galeottiana* Kunth

*Dioscorea convolvulacea* var. *glabra* (Hemsl.) Uline ex R.Knuth = *Dioscorea convolvulacea* Cham. & Schltdl. subsp. *convolvulacea*

*Dioscorea convolvulacea* var. *viridis* Uline = *Dioscorea galeottiana* Kunth

*Dioscorea cornifolia* Kunth = *Dioscorea pubera* Blume

*Dioscorea corumbensis* R.Knuth = *Dioscorea acanthogene* Rusby

*Dioscorea costaricensis* R.Knuth = *Dioscorea pilosiuscula* Bertero ex Spreng.

*Dioscorea costermansiana* De Wild. & T.Durand = *Dioscorea praehensilis* Benth.

*Dioscorea coxii* Matuda = *Dioscorea racemosa* (Klotzsch) Uline

*Dioscorea crenata* Vell. = *Dioscorea sinuata* Vell.

*Dioscorea crinita* Hook.f. = *Dioscorea quartiniana* A.Rich.

*Dioscorea crispata* Roxb. = *Dioscorea bulbifera* L.

*Dioscorea crumenigera* Mart. ex Griseb. = *Dioscorea amaranthoides* C.Presl

*Dioscorea cryptantha* Baker = *Dioscorea quartiniana* A.Rich.

*Dioscorea cubijensis* R.Knuth = (Menispermaceae)

*Dioscorea cuspidata* Klotzsch ex Kunth = *Dioscorea amazonum* Mart. ex Griseb. var. *amazonum*

*Dioscorea cuspidata* Balb. ex Kunth = *Dioscorea pilosiuscula* Bertero ex Spreng.

*Dioscorea cyclophylla* Urb. = *Rajania cordata* L.

*Dioscorea cylindrica* Burm.f. = (Convolvulaceae)

*Dioscorea cylindrostachya* I.M.Johnst. = *Dioscorea fastigiata* Gay

*Dioscorea cymosula* var. *cinerea* Uline ex R.Knuth = *Dioscorea carpomaculata* var. *cinerea* (Uline ex R.Knuth) O.Téllez & B.G.Schub.

*Dioscorea cymosula* var. *duchassaingii* Uline ex R.Knuth = *Dioscorea cymosula* Hemsl.

*Dioscorea cymosula* var. *longiracemosa* Uline ex R.Knuth = *Dioscorea cymosula* Hemsl.

*Dioscorea cynanchifolia* Griseb. = *Dioscorea marginata* Griseb.

*Dioscorea daemona* Roxb. = *Dioscorea hispida* Dennst.

*Dioscorea daemona* var. *reticulata* Hook.f. = *Dioscorea hispida* Dennst.

*Dioscorea dallmannensis* Hatus. = [43 NWG]

*Dioscorea dawei* De Wild. = *Dioscorea preussii* Pax subsp. *preussii*

*Dioscorea deamii* Matuda = *Dioscorea mexicana* Scheidw.

*Dioscorea debilis* var. *sagittifolia* Uline ex R.Knuth = *Dioscorea debilis* Uline ex R.Knuth

*Dioscorea decaisneana* Carrière = *Dioscorea polystachya* Turcz.

*Dioscorea decipiens* var. *glabrescens* C.T.Ting & M.C.Chang = *Dioscorea decipiens* Hook.f.

*Dioscorea deflexa* Hook.f. = *Dioscorea prainiana* R.Knuth

*Dioscorea dejantiana* Lem. = ?

*Dioscorea deltoidea* var. *orbiculata* Prain & Burkill = *Dioscorea deltoidea* Wall. ex Griseb.

*Dioscorea deltoidea* var. *sikkimensis* Prain = *Dioscorea prazeri* Prain & Burkill

*Dioscorea demeusei* De Wild. & T.Durand = *Dioscorea smilacifolia* De Wild. & T.Durand

*Dioscorea deppei* Schiede ex Schltdl. = *Dioscorea mexicana* Scheidw.

*Dioscorea diepenhorstiana* Miq. = *Dioscorea pyrifolia* Kunth var. *pyrifolia*

*Dioscorea digitaria* R.Knuth = *Dioscorea multiloba* Kunth

*Dioscorea digitata* Mill. = *Dioscorea pentaphylla* L.

*Dioscorea dinteri* Schinz = *Dioscorea quartiniana* A.Rich.

*Dioscorea discolor* Kunth = *Dioscorea dodecaneura* Vell.

*Dioscorea dissecta* R.Knuth = *Dioscorea kamoonensis* Kunth

*Dioscorea dodecandra* Steud. = *Dioscorea dodecaneura* Vell.

*Dioscorea dodecaneura* var. *maronensis* Uline ex R.Knuth = *Dioscorea dodecaneura* Vell.

*Dioscorea dodecaneura* var. *villosa* R.Knuth = *Dioscorea dodecaneura* Vell.

*Dioscorea dodecasemina* Caddick & Wilkin = *Dioscorea gaumeri* R.Knuth

*Dioscorea doryphora* Hance = *Dioscorea polystachya* Turcz.

*Dioscorea dusenii* Uline ex R.Knuth = *Dioscorea hirtiflora* Benth. subsp. *hirtiflora*

*Dioscorea ealensis* De Wild. = *Dioscorea minutiflora* Engl.

*Dioscorea eburina* Lour. = *Dioscorea alata* L.

*Dioscorea eburnea* Lour. = *Dioscorea alata* L.

*Dioscorea echinata* R.Knuth = ***Dioscorea cumingii*** Prain & Burkill

*Dioscorea echinulata* De Wild. = ***Dioscorea smilacifolia*** De Wild. & T.Durand

*Dioscorea effusa* Griseb. = ***Dioscorea fodinarum*** Kunth

*Dioscorea ekolo* De Wild. = ***Dioscorea minutiflora*** Engl.

*Dioscorea eldorado* Linden & André = ***Dioscorea amaranthoides*** C.Presl

*Dioscorea elegans* R.Knuth = ***Dioscorea schubertiae*** Ayala

*Dioscorea elegantula* Kunth = ***Dioscorea amazonum*** Mart. ex Griseb. var. ***amazonum***

*Dioscorea elephantopus* Spreng. = ***Dioscorea elephantipes*** (L'Hér.) Engl.

*Dioscorea elmeri* Prain & Burkill = ***Dioscorea cumingii*** Prain & Burkill

*Dioscorea engbo* De Wild. = ***Dioscorea minutiflora*** Engl.

*Dioscorea engleriana* R.Knuth = ***Dioscorea delavayi*** Franch.

*Dioscorea enneaneura* (Uline ex Diels) Prain & Burkill = ***Dioscorea tokoro*** Makino ex Miyabe

*Dioscorea entomophila* var. *tomentosa* Hauman = ***Dioscorea entomophila*** Hauman

*Dioscorea esculenta* var. *fasciculata* (Roxb.) Prain & Burkill = ***Dioscorea esculenta*** (Lour.) Burkill

*Dioscorea esculenta* var. *fulvidotomentosa* R.Knuth = ***Dioscorea esculenta*** (Lour.) Burkill

*Dioscorea esculenta* var. *spinosa* (Roxb. ex Prain & Burkill) R.Knuth = ***Dioscorea esculenta*** (Lour.) Burkill

*Dioscorea esculenta* var. *tiliifolia* (Kunth) Fosberg & Sachet = ***Dioscorea esculenta*** (Lour.) Burkill

*Dioscorea esurientium* Uline = ***Dioscorea convolvulacea*** Cham. & Schltdl. subsp. ***convolvulacea***

*Dioscorea excisa* R.Knuth = ***Dioscorea quartiniana*** A.Rich.

*Dioscorea fargesii* Franch. = ***Dioscorea kamoonensis*** Kunth

*Dioscorea fasciculata* Roxb. = ***Dioscorea esculenta*** (Lour.) Burkill

*Dioscorea fauriei* R.Knuth = ***Dioscorea japonica*** Thunb. var. ***japonica***

*Dioscorea ferruginea* Thunb. ex Prain & Burkill = ***Dioscorea pyrifolia*** Kunth var. ***pyrifolia***

*Dioscorea ferruginicaulis* Rusby = ***Dioscorea stegelmanniana*** R.Knuth

*Dioscorea filiformis* Griseb. = ***Dioscorea grisebachii*** Kunth

*Dioscorea filipendula* Dombey ex Kunth = ***Dioscorea humifusa*** Poepp.

*Dioscorea filirachis* R.Knuth = ***Dioscorea microbotrya*** Griseb.

*Dioscorea fimbriata* Uline ex R.Knuth = ***Dioscorea caldasensis*** R.Knuth

*Dioscorea firma* R.Knuth = ***Dioscorea kamoonensis*** Kunth

*Dioscorea flamignii* De Wild. = ***Dioscorea smilacifolia*** De Wild. & T.Durand

*Dioscorea fluminensis* R.Knuth = ***Dioscorea martiana*** Griseb.

*Dioscorea forbesii* Baker = ***Dioscorea quartiniana*** A.Rich.

*Dioscorea foxworthyi* Prain & Burkill = ***Dioscorea divaricata*** Blanco

*Dioscorea fracta* Griseb. = ***Dioscorea furcata*** Griseb.

*Dioscorea friedrichsthalii* R.Knuth = ***Dioscorea spiculiflora*** Hemsl.

*Dioscorea friesii* R.Knuth = ***Dioscorea monadelpha*** (Kunth) Griseb.

*Dioscorea frutescens* Rusby = ***Dioscorea coriacea*** Humb. & Bonpl. ex Willd.

*Dioscorea fulvida* Stapf = ***Dioscorea schimperiana*** Hochst. ex Kunth

*Dioscorea galipanensis* Klotzsch ex Kunth = ***Dioscorea trifoliata*** Kunth

*Dioscorea gayi* Phil. = ***Dioscorea fastigiata*** Gay

*Dioscorea gedensis* Prain & Burkill = ***Dioscorea madiunensis*** Prain & Burkill

*Dioscorea geissei* Phil. = ***Dioscorea fastigiata*** Gay

*Dioscorea georgensis* R.Knuth = ***Dioscorea amazonum*** Mart. ex Griseb. var. ***amazonum***

*Dioscorea gibbiflora* Hook.f. = ***Dioscorea filiformis*** Blume

*Dioscorea gibertii* R.Knuth = ***Dioscorea microbotrya*** Griseb.

*Dioscorea giraldii* R.Knuth = ***Dioscorea nipponica*** Makino

*Dioscorea glabra* var. *hastifolia* Prain & Burkill = ***Dioscorea glabra*** Roxb.

*Dioscorea glabra* var. *vera* Prain & Burkill = ***Dioscorea glabra*** Roxb.

*Dioscorea glauca* Rusby = ***Dioscorea coripatenis*** J.F.Macbr.

*Dioscorea glauca* Muhl. ex L.C.Beck = ***Dioscorea villosa*** L.

*Dioscorea glaucoidea* R.Knuth = ***Dioscorea nummularia*** Lam.

*Dioscorea globifera* R.Knuth = ***Dioscorea pentaphylla*** L.

*Dioscorea globosa* Roxb. = ***Dioscorea alata*** L.

*Dioscorea glomerulata* var. *mandoni* (Uline ex R.Knuth) R.Knuth = ***Dioscorea glomerulata*** Hauman

*Dioscorea goeringiana* Kunth = ***Dioscorea japonica*** Thunb. var. ***japonica***

*Dioscorea gossweileri* R.Knuth = ***Dioscorea quartiniana*** A.Rich.

*Dioscorea gouanioides* (Chodat & Hassl.) R.Knuth = ***Dioscorea acanthogene*** Rusby

*Dioscorea goyazensis* Griseb. = ***Dioscorea trifida*** L.f.

*Dioscorea gracillima* Ridl. = ***Dioscorea salicifolia*** Blume

*Dioscorea gracillima* var. *collettii* (Hook.f.) Uline ex R.Knuth = ***Dioscorea collettii*** Hook.f.

*Dioscorea grandibulbosa* R.Knuth = ***Dioscorea minutiflora*** Engl.

*Dioscorea grandiflora* M.Martens & Galeotti = ***Dioscorea galeottiana*** Kunth

*Dioscorea grandifolia* Schltdl. = ***Dioscorea convolvulacea*** subsp. ***grandifolia*** (Schltdl.) Uline ex R.Knuth

*Dioscorea grisebachii* Britton ex Léon = ***Dioscorea nipensis*** R.A.Howard

*Dioscorea guanaiensis* R.Knuth = ***Dioscorea acanthogene*** Rusby

*Dioscorea guaranitica* Chodat & Hassl. = ***Dioscorea subhastata*** Vell.

*Dioscorea guaranitica* var. *balansae* Pellegr. = ***Dioscorea subhastata*** Vell.

*Dioscorea guaranitica* f. *membranacea* Chodat & Hassl. = ***Dioscorea subhastata*** Vell.

*Dioscorea guaranitica* f. *subcoriacea* Chodat & Hassl. = ***Dioscorea subhastata*** Vell.

*Dioscorea hainanensis* Prain & Burkill = ***Dioscorea fordii*** Prain & Burkill

*Dioscorea haitiensis* R.Knuth = ***Rajania ovata*** Sw.

*Dioscorea harrissii* R.Knuth = ***Dioscorea kingii*** R.Knuth

*Dioscorea hassleriana* var. *triloba* Chodat & Hassl. = ***Dioscorea hassleriana*** Chodat

*Dioscorea hastata* Vell. = ***Dioscorea hassleriana*** Chodat

*Dioscorea hastata* C.Presl = ***Dioscorea preslii*** Steud.

*Dioscorea hastata* var. *balansae* Uline ex R.Knuth = ***Dioscorea hassleriana*** Chodat

*Dioscorea hastata* var. *hassleriana* (Chodat) Uline ex R.Knuth = ***Dioscorea hassleriana*** Chodat

*Dioscorea hastata* var. *mattogrossensis* Uline ex R.Knuth = ***Dioscorea hassleriana*** Chodat

*Dioscorea hebantha* Mart. ex Griseb. = ***Dioscorea dodecaneura*** Vell.

*Dioscorea hederacea* Miers = ***Dioscorea bryoniifolia*** Poepp.

*Dioscorea hederifolia* Griseb. = ***Dioscorea sylvatica*** Eckl.

*Dioscorea helicifolia* Kunth = ***Dioscorea auriculata*** Poepp.

*Dioscorea helmiicarpa* R.Knuth = ***Dioscorea larecajensis*** Uline ex R.Knuth

*Dioscorea henryi* (Prain & Burkill) C.T.Ting = ***Dioscorea delavayi*** Franch.

*Dioscorea henryi* Uline ex Diels = ***Dioscorea zingiberensis*** C.H.Wright

*Dioscorea heptaneura* f. *latisinuata* Uline ex R.Knuth = ***Dioscorea heptaneura*** Vell.

*Dioscorea heptaneura* f. *tenuicaulis* Uline ex R.Knuth = ***Dioscorea diamantinensis*** R.Knuth

*Dioscorea heptaphylla* Sasaki = ***Dioscorea cumingii*** Prain & Burkill

*Dioscorea herradurensis* (R.Knuth) P.Wilson ex Alain = ***Dioscorea wrightii*** Uline ex R.Knuth

*Dioscorea heterophylla* Roxb. = ***Dioscorea bulbifera*** L.

*Dioscorea heterophylla* Poepp. = ***Dioscorea saxatilis*** Poepp.

*Dioscorea heterophylla* var. *longifolia* L.E.Navas = ***Dioscorea saxatilis*** Poepp.

*Dioscorea hexaphylla* Raf. = ***Dioscorea villosa*** L.

*Dioscorea hirsuta* M.Martens & Galeotti = ***Dioscorea convolvulacea*** Cham. & Schltdl. subsp. ***convolvulacea***

*Dioscorea hirsuta* Blume = ***Dioscorea hispida*** Dennst.

*Dioscorea hirsuta* var. *glabra* Uline = ***Dioscorea convolvulacea*** Cham. & Schltdl. subsp. ***convolvulacea***

*Dioscorea hirsuticaulis* C.B.Rob. = ***Dioscorea jaliscana*** S.Watson

*Dioscorea hirticaulis* Bartlett = ***Dioscorea villosa*** L.

*Dioscorea hispida* var. *daemona* (Roxb.) Prain & Burkill = ***Dioscorea hispida*** Dennst.

*Dioscorea hispida* var. *reticulata* (Hook.f.) Sanjappa = ***Dioscorea hispida*** Dennst.

*Dioscorea hitchcockii* R.Knuth = ***Dioscorea amazonum*** Mart. ex Griseb. var. ***amazonum***

*Dioscorea hockii* De Wild. = ***Dioscorea schimperiana*** Hochst. ex Kunth

*Dioscorea hoehneana* R.Knuth = ***Dioscorea altissima*** Lam.

*Dioscorea hoffa* Cordem. = ***Dioscorea bulbifera*** L.

*Dioscorea hofika* Jum. & H.Perrier = ***Dioscorea bulbifera*** L.

*Dioscorea holstii* Harms = ***Dioscorea quartiniana*** A.Rich.

*Dioscorea holtii* R.Knuth = ***Dioscorea amazonum*** Mart. ex Griseb. var. ***amazonum***

*Dioscorea hongkongensis* Uline ex R.Knuth = ***Dioscorea glabra*** Roxb.

*Dioscorea huallagensis* R.Knuth = ***Dioscorea dodecaneura*** Vell.

*Dioscorea huberi* R.Knuth = ***Dioscorea amazonum*** Mart. ex Griseb. var. ***amazonum***

*Dioscorea hui* R.Knuth = ***Dioscorea collettii*** Hook.f. var. ***collettii***

*Dioscorea humblotii* R.Knuth = ***Dioscorea arcuatinervis*** Hochr.

*Dioscorea humifusa* var. *gracilis* (Hook. ex Poepp.) L.E.Navas = ***Dioscorea gracilis*** Hook. ex Poepp.

*Dioscorea* ×*hybrida* G.Nicholson = *D. communis* × *D. polystachya*

*Dioscorea hylophila* Harms = ***Dioscorea preussii*** subsp. ***hylophila*** (Harms) Wilkin

*Dioscorea hypoglauca* Palib. = ***Dioscorea collettii*** var. ***hypoglauca*** (Palib.) S.J.Pei & C.T.Ting

*Dioscorea hypotricha* Uline = ***Dioscorea semperflorens*** Uline

*Dioscorea hystrix* R.Knuth = ***Dioscorea minutiflora*** Engl.

*Dioscorea illustrata* W.Bull = ***Dioscorea dodecaneura*** Vell.

*Dioscorea inaequifolia* Elmer ex Prain & Burkill = ***Dioscorea cumingii*** Prain & Burkill

*Dioscorea izuensis* Akahori = ***Dioscorea collettii*** var. ***hypoglauca*** (Palib.) S.J.Pei & C.T.Ting

*Dioscorea jacquemontii* Hook.f. = ***Dioscorea pentaphylla*** L.

*Dioscorea japonica* var. *kelungensis* Prain & Burkill = ***Dioscorea japonica*** Thunb. var. ***japonica***

*Dioscorea japonica* var. *pseudojaponica* (Hayata) Yamam. = ***Dioscorea japonica*** Thunb. var. ***japonica***

*Dioscorea japonica* var. *vera* Prain & Burkill = ***Dioscorea japonica*** Thunb. var. ***japonica***

*Dioscorea javanica* Queva = ***Dioscorea alata*** L.

*Dioscorea joseensis* R.Knuth = ***Dioscorea convolvulacea*** Cham. & Schltdl. subsp. ***convolvulacea***

*Dioscorea junodii* Burtt Davy = ***Dioscorea sylvatica*** Eckl.

*Dioscorea kamoonensis* var. *brevifolia* Prain & Burkill = ***Dioscorea kamoonensis*** Kunth

*Dioscorea kamoonensis* var. *delavayi* (Franch.) Prain & Burkill = ***Dioscorea delavayi*** Franch.

*Dioscorea kamoonensis* var. *fargesii* (Franch.) Prain & Burkill = ***Dioscorea kamoonensis*** Kunth

*Dioscorea kamoonensis* var. *henryi* Prain & Burkill = ***Dioscorea delavayi*** Franch.

*Dioscorea kamoonensis* var. *praecox* Prain & Burkill = ***Dioscorea ochroleuca*** K.Y.Guan & D.F.Chamb.

*Dioscorea kamoonensis* var. *straminea* Prain & Burkill = ***Dioscorea kamoonensis*** Kunth

*Dioscorea kamoonensis* var. *vera* Prain & Burkill = ***Dioscorea kamoonensis*** Kunth

*Dioscorea kaoi* T.S.Liu & T.C.Huang = ***Dioscorea collettii*** var. ***hypoglauca*** (Palib.) S.J.Pei & C.T.Ting

*Dioscorea kegeliana* Griseb. = ***Dioscorea polygonoides*** Humb. & Bonpl. ex Willd.

*Dioscorea kelungensis* Hayata = ***Dioscorea collettii*** Hook.f. var. ***collettii***

*Dioscorea kelungensis* R.Knuth = ***Dioscorea japonica*** Thunb. var. ***japonica***

*Dioscorea kiangsiensis* R.Knuth = ***Dioscorea japonica*** Thunb. var. ***japonica***

*Dioscorea kita* Queva = ***Dioscorea dodecaneura*** Vell.

*Dioscorea kleiniana* Kunth = ***Dioscorea pentaphylla*** L.

*Dioscorea klugii* R.Knuth = ***Dioscorea amazonum*** var. ***klugii*** (R.Knuth) Ayala

*Dioscorea knuthii* H.Perrier = ***Dioscorea tsaratananensis*** H.Perrier

*Dioscorea koordersii* R.Knuth = ***Dioscorea filiformis*** Blume

*Dioscorea korrorensis* R.Knuth = ***Dioscorea bulbifera*** L.

*Dioscorea kunthiana* Uline ex R.Knuth = ***Dioscorea grisebachii*** Kunth

*Dioscorea kweichowensis* R.Knuth = [36 CHC]

*Dioscorea lagoa-santa* Uline ex R.Knuth = ***Dioscorea subhastata*** Vell.

*Dioscorea lanceolata* C.Wright ex Griseb. = *Rajania quinquefolia* L.

*Dioscorea lanosa* Gleason = **Dioscorea pilosiuscula** Bertero ex Spreng.

*Dioscorea latifolia* Benth. = **Dioscorea bulbifera** L.

*Dioscorea laurifolia* var. *hookeri* R.Knuth = **Dioscorea laurifolia** Wall. ex Hook.f.

*Dioscorea laxiflora* Schltdl. = **Dioscorea remotiflora** Kunth

*Dioscorea laxiflora* var. *auriculata* Griseb. = **Dioscorea laxiflora** Mart. ex Griseb.

*Dioscorea laxiflora* var. *calystegioides* (Kunth) Uline ex R.Knuth = **Dioscorea laxiflora** Mart. ex Griseb.

*Dioscorea laxiflora* var. *cincinnata* Uline ex R.Knuth = **Dioscorea laxiflora** Mart. ex Griseb.

*Dioscorea laxiflora* var. *cissifolia* Uline ex R.Knuth = **Dioscorea laxiflora** Mart. ex Griseb.

*Dioscorea laxiflora* var. *truncata* Griseb. = **Dioscorea laxiflora** Mart. ex Griseb.

*Dioscorea laxiflora* var. *truncatolanceolata* Uline ex R.Knuth = **Dioscorea laxiflora** Mart. ex Griseb.

*Dioscorea lecardii* De Wild. = **Dioscorea sagittifolia** var. **lecardii** (De Wild.) Nkounkou

*Dioscorea ledermannii* R.Knuth = **Dioscorea flabellifolia** Prain & Burkill

*Dioscorea leiboldiana* Kunth = **Dioscorea mexicana** Scheidw.

*Dioscorea lepcharum* var. *bhamoica* Prain & Burkill = **Dioscorea lepcharum** Prain & Burkill

*Dioscorea leptostachya* Gardner = **Dioscorea martiana** Griseb.

*Dioscorea liebrechtsiana* De Wild. & T.Durand = **Dioscorea praehensilis** Benth.

*Dioscorea lilela* De Wild. = **Dioscorea minutiflora** Engl.

*Dioscorea lilloi* (Hauman) Hauman = **Dioscorea megalantha** Griseb.

*Dioscorea lindiensis* R.Knuth = **Dioscorea hirtiflora** subsp. **orientalis** Milne-Redh.

*Dioscorea lindmanii* Uline ex R.Knuth = **Dioscorea pilosiuscula** Bertero ex Spreng.

*Dioscorea linearis* Griseb. = **Dioscorea nipensis** R.A.Howard

*Dioscorea linearis* Bertero ex Colla = **Dioscorea saxatilis** Poepp.

*Dioscorea litoie* De Wild. = **Dioscorea minutiflora** Engl.

*Dioscorea lloydii* E.H.L.Krause = **Dioscorea villosa** L.

*Dioscorea lobata* Uline = **Dioscorea galeottiana** Kunth

*Dioscorea lobata* var. *lasiophylla* Uline ex R.Knuth = **Dioscorea galeottiana** Kunth

*Dioscorea lobata* var. *morelosana* Uline = **Dioscorea morelosana** (Uline) Matuda

*Dioscorea longifolia* Phil. = **Dioscorea saxatilis** Poepp.

*Dioscorea longifolia* Raf. = **Dioscorea villosa** L.

*Dioscorea longipetiolata* Baudon = **Dioscorea bulbifera** L.

*Dioscorea longirachis* R.Knuth = **Dioscorea monadelpha** (Kunth) Griseb.

*Dioscorea longispicata* De Wild. = **Dioscorea preussii** Pax subsp. **preussii**

*Dioscorea lorentzii* Uline ex R.Knuth = **Dioscorea glomerulata** Hauman

*Dioscorea lorentzii* var. *mandonii* Uline ex R.Knuth = **Dioscorea glomerulata** Hauman

*Dioscorea lucida* Scott-Elliot = **Dioscorea bemarivensis** Jum. & H.Perrier

*Dioscorea lufensis* R.Knuth = **Dioscorea nummularia** Lam.

*Dioscorea lunata* Roth = **Dioscorea hispida** Dennst.

*Dioscorea luschnathiana* Kunth = **Dioscorea pohlii** var. **luschnathiana** (Kunth) Uline ex R.Knuth

*Dioscorea lutea* G.Mey. = **Dioscorea polygonoides** Humb. & Bonpl. ex Willd.

*Dioscorea macabiha* Jum. & H.Perrier = **Dioscorea sansibarensis** Pax

*Dioscorea maciba* Jum. & H.Perrier = **Dioscorea sansibarensis** Pax

*Dioscorea macrocapsa* Uline = **Dioscorea multiflora** Mart. ex Griseb.

*Dioscorea macrophylla* M.Martens & Galeotti = **Dioscorea mexicana** Scheidw.

*Dioscorea macrostachya* M.Martens & Galeotti = **Dioscorea convolvulacea** Cham. & Schltdl. subsp. **convolvulacea**

*Dioscorea macrostachya* Benth. = **Dioscorea mexicana** Scheidw.

*Dioscorea macrostachya* var. *palmeri* (R.Knuth) C.V.Morton = **Dioscorea palmeri** R.Knuth

*Dioscorea macrostachya* var. *sessiliflora* Uline = **Dioscorea mexicana** Scheidw.

*Dioscorea macroura* Harms = **Dioscorea sansibarensis** Pax

*Dioscorea madagascariensis* R.Knuth = **Dioscorea bemarivensis** Jum. & H.Perrier

*Dioscorea mairei* H.Lév. = **Dioscorea hemsleyi** Prain & Burkill

*Dioscorea mairei* R.Knuth = **Dioscorea kamoonensis** Kunth

*Dioscorea majungensis* R.Knuth = **Dioscorea bemarivensis** Jum. & H.Perrier

*Dioscorea malchairii* De Wild. = **Dioscorea preussii** Pax subsp. **preussii**

*Dioscorea malifolia* Baker = **Dioscorea cotinifolia** Kunth

*Dioscorea maliliensis* R.Knuth = **Dioscorea prainiana** R.Knuth

*Dioscorea maranonensis* R.Knuth = **Dioscorea altissima** Lam.

*Dioscorea marlothii* R.Knuth = **Dioscorea sylvatica** Eckl.

*Dioscorea martiana* var. *caudata* R.Knuth = **Dioscorea martiana** Griseb.

*Dioscorea martiana* var. *genuina* R.Knuth = **Dioscorea martiana** Griseb.

*Dioscorea martiana* var. *leptostachya* (Gardner) Uline ex R.Knuth = **Dioscorea martiana** Griseb.

*Dioscorea martiana* var. *pedicellata* R.Knuth = **Dioscorea martiana** Griseb.

*Dioscorea martinicensis* Spreng. = **Dioscorea polygonoides** Humb. & Bonpl. ex Willd.

*Dioscorea matsudae* Hayata = **Dioscorea cirrhosa** Lour. var. **cirrhosa**

*Dioscorea maximowiczii* Uline ex R.Knuth = **Dioscorea tenuipes** Franch. & Sav.

*Dioscorea maynensis* Kunth = **Dioscorea piperifolia** Humb. & Bonpl. ex Willd.

*Dioscorea megalantha* var. *lilloi* Hauman = **Dioscorea megalantha** Griseb.

*Dioscorea megalantha* var. *subsessilis* Hauman = **Dioscorea hieronymi** Uline ex R.Knuth

*Dioscorea megalobotrya* Kunth & R.H.Schomb. = **Dioscorea amazonum** Mart. ex Griseb. var. **amazonum**

*Dioscorea megaptera* Raf. = **Dioscorea villosa** L.

*Dioscorea mengtzeana* R.Knuth = **Dioscorea kamoonensis** Kunth

*Dioscorea mexicana* var. *sessiliflora* (Uline) Matuda = **Dioscorea mexicana** Scheidw.

*Dioscorea micrantha* Kunth = **Dioscorea martiana** Griseb.

*Dioscorea microbotrya* var. *grandifolia* Hauman = **Dioscorea microbotrya** Griseb.

*Dioscorea microcuspis* Baker = **Dioscorea retusa** Mast.

*Dioscorea microphylla* Steud. = **Dioscorea variifolia** Bertero

*Dioscorea mildbraediana* R.Knuth = **Dioscorea buchananii** Benth.

*Dioscorea mildbraedii* R.Knuth = **Dioscorea sagittifolia** var. **lecardii** (De Wild.) Nkounkou

*Dioscorea mollis* var. *pachycarpa* (Kunth) Uline ex R.Knuth = **Dioscorea mollis** Kunth

*Dioscorea mollissima* Blume = **Dioscorea hispida** Dennst.

*Dioscorea moma* De Wild. = **Dioscorea cayennensis** Lam. subsp. **cayennensis**

*Dioscorea monadelpha* var. *opaca* Hicken = **Dioscorea subhastata** Vell.

*Dioscorea monadelphoides* J.F.Macbr. = **Dioscorea monadelpha** (Kunth) Griseb.

*Dioscorea monadelphoides* var. *longirachis* (R.Knuth) Ayala = **Dioscorea monadelpha** (Kunth) Griseb.

*Dioscorea montana* (Burch.) Spreng. = **Dioscorea elephantipes** (L'Hér.) Engl.

*Dioscorea montana* var. *duemmeri* R.Knuth = **Dioscorea sylvatica** Eckl.

*Dioscorea montana* var. *glauca* R.Knuth = **Dioscorea sylvatica** Eckl.

*Dioscorea montana* var. *paniculata* (Kuntze) R.Knuth = **Dioscorea sylvatica** Eckl.

*Dioscorea montecristina* Hadac = **Dioscorea nipensis** R.A.Howard

*Dioscorea morobeensis* R.Knuth = **Dioscorea opaca** R.Knuth

*Dioscorea morsei* Prain & Burkill = **Dioscorea collettii** var. **hypoglauca** (Palib.) S.J.Pei & C.T.Ting

*Dioscorea multicolor* Linden & André = **Dioscorea amaranthoides** C.Presl

*Dioscorea multicolor* var. *chrysophylla* Linden & André = **Dioscorea amaranthoides** C.Presl

*Dioscorea multicolor* var. *melanoleuca* Linden & André = **Dioscorea amaranthoides** C.Presl

*Dioscorea multicolor* var. *metallica* Linden & André = **Dioscorea amaranthoides** C.Presl

*Dioscorea multicolor* var. *sagittaria* Linden & André = **Dioscorea amaranthoides** C.Presl

*Dioscorea multiflora* Engl. ex Pax = **Dioscorea minutiflora** Engl.

*Dioscorea multiflora* C.Presl = **Dioscorea polygonoides** Humb. & Bonpl. ex Willd.

*Dioscorea multiflora* var. *asuncianensis* Uline ex R.Knuth = **Dioscorea multiflora** Mart. ex Griseb.

*Dioscorea multiflora* var. *asuncionensis* Uline = **Dioscorea multiflora** Mart. ex Griseb.

*Dioscorea multiflora* var. *concepcionis* Pellegr. = **Dioscorea multiflora** Mart. ex Griseb.

*Dioscorea multiflora* var. *gouanioides* Chodat & Hassl. = **Dioscorea acanthogene** Rusby

*Dioscorea multiflora* var. *grandifolia* Griseb. = **Dioscorea fodinarum** Kunth

*Dioscorea multiflora* var. *loefgrenii* Uline ex R.Knuth = **Dioscorea multiflora** Mart. ex Griseb.

*Dioscorea multiflora* var. *lofgrenii* R.Knuth = **Dioscorea multiflora** Mart. ex Griseb.

*Dioscorea multiflora* var. *parvifolia* Griseb. = **Dioscorea multiflora** Mart. ex Griseb.

*Dioscorea multispicata* R.Knuth = **Dioscorea glandulosa** (Griseb.) Klotzsch ex Kunth var. **glandulosa**

*Dioscorea myriantha* Kunth = **Dioscorea filiformis** Blume

*Dioscorea nana* Schltdl. = **Dioscorea multinervis** Benth.

*Dioscorea narinensis* R.Knuth = **Dioscorea acanthogene** Rusby

*Dioscorea neglecta* R.Knuth = **Dioscorea japonica** Thunb. var. *japonica*

*Dioscorea nepalensis* Sweet ex Bernardi = **Dioscorea deltoidea** Wall. ex Griseb.

*Dioscorea nesiotis* Hemsl. = **Dioscorea bemarivensis** Jum. & H.Perrier

*Dioscorea niederleinii* R.Knuth = **Dioscorea multiflora** Mart. ex Griseb.

*Dioscorea nigrescens* R.Knuth = **Dioscorea collettii** Hook.f. var. **collettii**

*Dioscorea nigrescens* Phil. = **Dioscorea humifusa** Poepp.

*Dioscorea nipponica* var. *jamesii* Prain & Burkill = **Dioscorea nipponica** Makino

*Dioscorea nipponica* var. *rosthornii* (Diels) Prain & Burkill = **Dioscorea nipponica** Makino

*Dioscorea nipponica* subsp. *rosthornii* (Diels) C.T.Ting = **Dioscorea nipponica** Makino

*Dioscorea nitida* R.Knuth = **Dioscorea altissima** Lam.

*Dioscorea nodosa* R.Knuth = **Dioscorea glomerulata** Hauman

*Dioscorea novemloba* Steud. ex F.Phil. = **Dioscorea brachybotrya** Poepp.

*Dioscorea nummularia* var. *belophylla* Prain = **Dioscorea belophylla** (Prain) Voigt ex Haines

*Dioscorea nummularia* var. *glauca* Prain & Burkill = **Dioscorea spicata** Roth

*Dioscorea nummularia* var. *lata* R.Knuth = **Dioscorea nummularia** Lam.

*Dioscorea nurii* R.Knuth = **Dioscorea kingii** R.Knuth

*Dioscorea obtusifolia* var. *philippii* Uline ex R.Knuth = **Dioscorea obtusifolia** Hook. & Arn.

*Dioscorea occidentalis* R.Knuth = **Dioscorea cayennensis** Lam. subsp. **cayennensis**

*Dioscorea odoratissima* Pax = **Dioscorea praehensilis** Benth.

*Dioscorea oenea* Prain & Burkill = **Dioscorea collettii** Hook.f. var. **collettii**

*Dioscorea opposita* Thunb. = **Dioscorea oppositifolia** L.

*Dioscorea oppositifolia* var. *dukhunensis* Prain & Burkill = **Dioscorea oppositifolia** L.

*Dioscorea oppositifolia* var. *linnaei* Prain & Burkill = **Dioscorea oppositifolia** L.

*Dioscorea oppositifolia* var. *thwaitesii* Prain & Burkill = **Dioscorea oppositifolia** L.

*Dioscorea orbicularis* A.Chev. ex De Wild. = **Dioscorea smilacifolia** De Wild. & T.Durand

*Dioscorea orthogoneura* var. *acutissima* Uline ex R.Knuth = **Dioscorea orthogoneura** Uline ex Hochr.

*Dioscorea orthogoneura* var. *brevispicata* Uline ex R.Knuth = **Dioscorea orthogoneura** Uline ex Hochr.

*Dioscorea orthogoneura* var. *meiapontensis* Uline ex R.Knuth = **Dioscorea orthogoneura** Uline ex Hochr.

*Dioscorea ovifotsy* H.Perrier = **Dioscorea seriflora** Jum. & H.Perrier

*Dioscorea owenii* Prain & Burkill = [42 MLY]

*Dioscorea oxyphylla* R.Knuth = **Dioscorea divaricata** Blanco

*Dioscorea pachycarpa* Kunth = **Dioscorea mollis** Kunth

*Dioscorea palauensis* R.Knuth = **Dioscorea nummularia** Lam.

*Dioscorea palmata* Juss. ex Pers. = **Dioscorea trifida** L.f.

*Dioscorea palopoensis* R.Knuth = **Dioscorea nummularia** Lam.

*Dioscorea paniculata* Michx. = **Dioscorea villosa** L.

*Dioscorea paniculata* var. *glabrifolia* Bartlett = **Dioscorea villosa** L.

*Dioscorea papillaris* Blanco = **Dioscorea esculenta** (Lour.) Burkill

*Dioscorea papuana* Ridl. = **Dioscorea elegans** Ridl. ex Prain & Burkill

*Dioscorea papuana* Warb. = **Dioscorea esculenta** (Lour.) Burkill

*Dioscorea paraguayensis* R.Knuth = **Dioscorea acanthogene** Rusby

*Dioscorea paranensis* R.Knuth = **Dioscorea glandulosa** (Griseb.) Klotzsch ex Kunth var. **glandulosa**

*Dioscorea parviflora* Phil. = **Dioscorea saxatilis** Poepp.

*Dioscorea parviflora* C.T.Ting = **Dioscorea sinoparviflora** C.T.Ting, M.G.Gilbert & Turland

*Dioscorea parvifolia* Phil. = **Dioscorea modesta** Phil.

*Dioscorea paupera* Phil. = **Dioscorea fastigiata** Gay

*Dioscorea pedatifida* Phil. = **Dioscorea reticulata** Gay

*Dioscorea pedicellata* Morong = **Dioscorea pilcomayensis** Hauman

*Dioscorea pellegrinii* Hassl. ex R.Knuth = **Dioscorea pilcomayensis** Hauman

*Dioscorea peltata* Burm.f. = (Menispermaceae)

*Dioscorea pendula* R.Knuth = **Dioscorea minutiflora** Engl.

*Dioscorea pennellii* R.Knuth = **Dioscorea coriacea** Humb. & Bonpl. ex Willd.

*Dioscorea pennellii* var. *pilosula* R.Knuth = **Dioscorea coriacea** Humb. & Bonpl. ex Willd.

*Dioscorea pentadactyla* (Pax) Welw. = **Dioscorea quartiniana** A.Rich.

*Dioscorea pentaphylla* var. *cardonii* Prain & Burkill = **Dioscorea pentaphylla** L.

*Dioscorea pentaphylla* var. *communis* Prain & Burkill = **Dioscorea pentaphylla** L.

*Dioscorea pentaphylla* var. *hortorum* Prain & Burkill = **Dioscorea pentaphylla** L.

*Dioscorea pentaphylla* var. *jacquemontii* (Hook.f.) Prain & Burkill = **Dioscorea pentaphylla** L.

*Dioscorea pentaphylla* var. *kussok* Prain & Burkill = **Dioscorea pentaphylla** L.

*Dioscorea pentaphylla* var. *linnaei* Prain & Burkill = **Dioscorea pentaphylla** L.

*Dioscorea pentaphylla* var. *malaica* Prain & Burkill = **Dioscorea pentaphylla** L.

*Dioscorea pentaphylla* var. *papuana* Burkill = **Dioscorea pentaphylla** L.

*Dioscorea pentaphylla* var. *rheedei* Prain & Burkill = **Dioscorea pentaphylla** L.

*Dioscorea pentaphylla* var. *simplicifolia* Prain & Burkill = **Dioscorea pentaphylla** L.

*Dioscorea pentaphylla* var. *suli* Prain & Burkill = **Dioscorea pentaphylla** L.

*Dioscorea pentaphylla* var. *thwaitesii* Prain & Burkill = **Dioscorea pentaphylla** L.

*Dioscorea pentaphylla* var. *unifoliata* R.Knuth = **Dioscorea pentaphylla** L.

*Dioscorea peperoides* var. *sagittifolia* Prain & Burkill = **Dioscorea peperoides** Prain & Burkill var. **peperoides**

*Dioscorea peperoides* var. *vera* Prain & Burkill = **Dioscorea peperoides** Prain & Burkill var. **peperoides**

*Dioscorea perforata* Bertero ex Spreng. = [81 LEE]

*Dioscorea permollis* R.Knuth = **Dioscorea cymosula** Hemsl.

*Dioscorea perrieri* R.Knuth = **Dioscorea bulbifera** L.

*Dioscorea persimilis* var. *pubescens* C.T.Ting & M.C.Chang = **Dioscorea persimilis** Prain & Burkill

*Dioscorea peteri* R.Knuth = **Dioscorea quartiniana** A.Rich.

*Dioscorea phaseoloides* Pax = **Dioscorea quartiniana** A.Rich.

*Dioscorea pichinchensis* R.Knuth = **Dioscorea cuspidata** Humb. & Bonpl. ex Willd.

*Dioscorea pilosiuscula* var. *panamensis* R.Knuth = **Dioscorea pilosiuscula** Bertero ex Spreng.

*Dioscorea piperifolia* Klotzsch ex Kunth = **Dioscorea polygonoides** Humb. & Bonpl. ex Willd.

*Dioscorea piperifolia* Griseb. = **Dioscorea polygonoides** Humb. & Bonpl. ex Willd.

*Dioscorea piperifolia* var. *apiculata* Uline = **Dioscorea piperifolia** Humb. & Bonpl. ex Willd.

*Dioscorea piperifolia* var. *glandulosa* Griseb. = **Dioscorea glandulosa** (Griseb.) Klotzsch ex Kunth

*Dioscorea piperifolia* var. *legitima* Griseb. = **Dioscorea glandulosa** (Griseb.) Klotzsch ex Kunth var. **glandulosa**

*Dioscorea piperifolia* var. *obtusifolia* Chodat & Hassl. = **Dioscorea piperifolia** Humb. & Bonpl. ex Willd.

*Dioscorea piperifolia* var. *triangularis* Griseb. = **Dioscorea glandulosa** (Griseb.) Klotzsch ex Kunth var. **glandulosa**

*Dioscorea piratinyensis* R.Knuth = **Dioscorea subhastata** Vell.

*Dioscorea pirita* Nadeaud = **Dioscorea nummularia** Lam.

*Dioscorea platanifolia* Prain & Burkill = **Dioscorea althaeoides** R.Knuth

*Dioscorea platystemon* Hauman = **Dioscorea hassleriana** Chodat

*Dioscorea poeppigii* Kunth = **Dioscorea altissima** Lam.

*Dioscorea polyantha* Rendle = **Dioscorea hirtiflora** Benth. subsp. **hirtiflora**

*Dioscorea polyclados* var. *oblongifolia* Uline ex R.Knuth = **Dioscorea polyclados** Hook.f.

*Dioscorea polygonoides* M.Martens & Galeotti = **Dioscorea pallens** Schltdl.

*Dioscorea polygonoides* var. *aperta* R.Knuth = **Dioscorea polygonoides** Humb. & Bonpl. ex Willd.

*Dioscorea polygonoides* var. *martinicensis* (Spreng.) R.Knuth = **Dioscorea polygonoides** Humb. & Bonpl. ex Willd.

*Dioscorea polygonoides* var. *scorpioidea* Uline = **Dioscorea polygonoides** Humb. & Bonpl. ex Willd.

*Dioscorea polygonoides* var. *sieberi* (Kunth) Uline = **Dioscorea polygonoides** Humb. & Bonpl. ex Willd.

*Dioscorea polyphylla* R.Knuth = **Dioscorea cumingii** Prain & Burkill

*Dioscorea porteri* Prain & Burkill ex Ridl. = **Dioscorea kingii** R.Knuth

*Dioscorea potaninii* Prain & Burkill = **Dioscorea polystachya** Turcz.

*Dioscorea pozucoensis* R.Knuth = **Dioscorea acanthogene** Rusby

*Dioscorea praecox* Prain & Burkill = **Dioscorea hemsleyi** Prain & Burkill

*Dioscorea praehensilis* var. *minutiflora* (Engl.) Baker = **Dioscorea minutiflora** Engl.

*Dioscorea praetervisa* R.Knuth = **Dioscorea coronata** Hauman

*Dioscorea preangeriana* Uline ex R.Knuth = **Dioscorea pyrifolia** Kunth var. **pyrifolia**

*Dioscorea prismatica* Linden & André = **Dioscorea amaranthoides** C.Presl

*Dioscorea propinqua* Hemsl. = **Dioscorea mexicana** Scheidw.

*Dioscorea pruinosa* A.Chev. = **Dioscorea cayennensis** Lam. subsp. **cayennensis**

*Dioscorea pruinosa* Kunth = **Dioscorea villosa** L.

*Dioscorea pseudobatatas* (Hauman) Herter = **Dioscorea polystachya** Turcz.

*Dioscorea pseudojaponica* Hayata = **Dioscorea japonica** Thunb. var. **japonica**

*Dioscorea pterocaulon* De Wild. & T.Durand = **Dioscorea preussii** Pax subsp. *preussii*

*Dioscorea pulchella* Roxb. = **Dioscorea bulbifera** L.

*Dioscorea pulverea* Prain & Burkill = **Dioscorea aspersa** Prain & Burkill

*Dioscorea pumila* Sessé & Moç. = **Dioscorea minima** C.B.Rob. & Seaton

*Dioscorea punctata* R.Br. = **Dioscorea transversa** R.Br.

*Dioscorea purpurea* Roxb. = **Dioscorea alata** L.

*Dioscorea pusilla* Hook. = **Dioscorea humilis** Bertero ex Colla

*Dioscorea pynaertioides* De Wild. = **Dioscorea minutiflora** Engl.

*Dioscorea quartiniana* var. *apiculata* (De Wild.) Burkill = **Dioscorea quartiniana** A.Rich.

*Dioscorea quartiniana* var. *cryptantha* (Baker) Burkill = **Dioscorea quartiniana** A.Rich.

*Dioscorea quartiniana* var. *dinteri* (Schinz) Burkill = **Dioscorea quartiniana** A.Rich.

*Dioscorea quartiniana* var. *excisa* (R.Knuth) Burkill = **Dioscorea quartiniana** A.Rich.

*Dioscorea quartiniana* var. *forbesii* (Baker) Burkill = **Dioscorea quartiniana** A.Rich.

*Dioscorea quartiniana* var. *hochstetteri* Engl. = **Dioscorea quartiniana** A.Rich.

*Dioscorea quartiniana* var. *holstii* (Harms) Burkill = **Dioscorea quartiniana** A.Rich.

*Dioscorea quartiniana* var. *latifolia* R.Knuth = **Dioscorea quartiniana** A.Rich.

*Dioscorea quartiniana* var. *pentadactyla* Pax = **Dioscorea quartiniana** A.Rich.

*Dioscorea quartiniana* var. *phaseoloides* (Pax) Burkill = **Dioscorea quartiniana** A.Rich.

*Dioscorea quartiniana* var. *schliebenii* Burkill = **Dioscorea quartiniana** A.Rich.

*Dioscorea quartiniana* var. *schweinfurthiana* (Pax) Burkill = **Dioscorea quartiniana** A.Rich.

*Dioscorea quartiniana* var. *stuhlmannii* (Harms) Burkill = **Dioscorea quartiniana** A.Rich.

*Dioscorea quartiniana* var. *subpedata* Chiov. = **Dioscorea quartiniana** A.Rich.

*Dioscorea quartiniana* var. *vestita* R.Knuth = **Dioscorea quartiniana** A.Rich.

*Dioscorea quaternata* Walter = **Dioscorea villosa** L.

*Dioscorea quaternata* var. *glauca* (Muhl. ex L.C.Beck) Fernald = **Dioscorea villosa** L.

*Dioscorea quinata* Walter = **Dioscorea villosa** L.

*Dioscorea quinquefoliolata* R.Knuth = **Dioscorea crotalariifolia** Uline

*Dioscorea quinqueloba* Thunb. = **Dioscorea quinquelobata** Thunb.

*Dioscorea quinquelobata* Vell. = **Dioscorea trifida** L.f.

*Dioscorea quirogae* R.Knuth = **Dioscorea ceratandra** Uline ex R.Knuth

*Dioscorea racemosa* Rusby = **Dioscorea dodecaneura** Vell.

*Dioscorea racemosa* var. *hoffmannii* Uline = **Dioscorea lepida** C.V.Morton

*Dioscorea raishaensis* Hayata = **Dioscorea persimilis** Prain & Burkill

*Dioscorea rajanioides* Uline ex R.Knuth = **Dioscorea altissima** Lam.

*Dioscorea ramonensis* R.Knuth = **Dioscorea macbrideana** R.Knuth

*Dioscorea rangunensis* R.Knuth = **Dioscorea birmanica** Prain & Burkill

*Dioscorea ravenii* Ayala = **Dioscorea nipensis** R.A.Howard

*Dioscorea raymundii* R.Knuth = **Dioscorea nummularia** Lam.

*Dioscorea recurva* Rusby = **Dioscorea glandulosa** (Griseb.) Klotzsch ex Kunth var. *glandulosa*

*Dioscorea rehmannii* Baker = **Dioscorea sylvatica** Eckl.

*Dioscorea remotiflora* var. *maculata* Uline = **Dioscorea remotiflora** Kunth

*Dioscorea remotiflora* var. *palmeri* Uline = **Dioscorea remotiflora** Kunth

*Dioscorea remotiflora* var. *sparsiflora* (Hemsl.) Uline = **Dioscorea remotiflora** Kunth

*Dioscorea repanda* Blume = **Dioscorea filiformis** Blume

*Dioscorea repanda* Raf. = **Dioscorea villosa** L.

*Dioscorea reticulata* var. *nervosa* (Phil.) L.E.Navas = **Dioscorea nervosa** Phil.

*Dioscorea revillae* Ayala = **Dioscorea altissima** Lam.

*Dioscorea rhacodes* Peter ex R.Knuth = **Dioscorea buchananii** Benth.

*Dioscorea rhipogonoides* Oliv. = **Dioscorea cirrhosa** Lour. var. *cirrhosa*

*Dioscorea riparia* Kunth & R.H.Schomb. = **Dioscorea altissima** Lam.

*Dioscorea rogersii* Prain & Burkill = **Dioscorea bulbifera** L.

*Dioscorea rosthornii* Diels = **Dioscorea polystachya** Turcz.

*Dioscorea rotundata* Poir. = **Dioscorea cayennensis** subsp. *rotundata* (Poir.) J.Miège

*Dioscorea rotundifoliolata* R.Knuth = **Dioscorea delavayi** Franch.

*Dioscorea rubella* Roxb. = **Dioscorea alata** L.

*Dioscorea rubiginosa* Benth. = **Dioscorea hirtiflora** Benth. subsp. *hirtiflora*

*Dioscorea rubricaulis* Kunth = **Dioscorea decorticans** C.Presl

*Dioscorea ruiziana* Klotzsch ex Kunth = **Dioscorea trifida** L.f.

*Dioscorea rumicoides* var. *longibracteata* R.Knuth = **Dioscorea rumicoides** Griseb.

*Dioscorea sagittata* Royle = **Dioscorea belophylla** (Prain) Voigt ex Haines

*Dioscorea saidae* R.Knuth = **Dioscorea tokoro** Makino ex Miyabe

*Dioscorea samydea* Griseb. = **Dioscorea altissima** Lam.

*Dioscorea samydea* var. *corcovadensis* Uline ex R.Knuth = **Dioscorea altissima** Lam.

*Dioscorea samydea* var. *poeppigii* (Kunth) Ayala = **Dioscorea altissima** Lam.

*Dioscorea sandakanensis* R.Knuth = **Dioscorea pyrifolia** Kunth var. *pyrifolia*

*Dioscorea sapindioides* C.Presl = **Dioscorea pilosiuscula** Bertero ex Spreng.

*Dioscorea sapinii* De Wild. = **Dioscorea alata** L.

*Dioscorea sarawakensis* R.Knuth = **Dioscorea salicifolia** Blume

*Dioscorea sativa* L. = **Dioscorea villosa** L.

*Dioscorea sativa* f. *domestica* Makino = **Dioscorea bulbifera** L.

*Dioscorea sativa* var. *elongata* F.M.Bailey = **Dioscorea bulbifera** L.

*Dioscorea sativa* var. *rotunda* F.M.Bailey = **Dioscorea bulbifera** L.

*Dioscorea scabra* var. *aspera* (Humb. & Bonpl. ex Willd.) Uline ex R.Knuth = **Dioscorea aspera** Humb. & Bonpl. ex Willd.

*Dioscorea scandens* Kunze ex Kunth = **Dioscorea brachybotrya** Poepp.

*Dioscorea schimperiana* var. *adamaowense* Jacq.-Fél. = **Dioscorea schimperiana** Hochst. ex Kunth

*Dioscorea schimperiana* var. *nigrescens* Uline ex R.Knuth = **Dioscorea longicuspis** R.Knuth

*Dioscorea schimperiana* var. *vestita* Pax = **Dioscorea schimperiana** Hochst. ex Kunth

*Dioscorea schlechtendalii* Kunth = **Dioscorea multinervis** Benth.

*Dioscorea schlechteri* Harms = **Dioscorea semperflorens** Uline

*Dioscorea schliebenii* R.Knuth = **Dioscorea quartiniana** A.Rich.

*Dioscorea schomburgkiana* (Kunth) Hochr. = **Dioscorea pilosiuscula** Bertero ex Spreng.

*Dioscorea schweinfurthiana* Pax = **Dioscorea quartiniana** A.Rich.

*Dioscorea scorpioidea* C.Wright = **Rajania cordata** L.

*Dioscorea scortechinii* var. *parviflora* Prain & Burkill = **Dioscorea scortechinii** Prain & Burkill

*Dioscorea seemannii* Prain & Burkill = **Dioscorea nummularia** Lam.

*Dioscorea sellowiana* var. *mantiqueirensis* R.Knuth = **Dioscorea sellowiana** Uline ex R.Knuth

*Dioscorea seniavinii* Prain & Burkill = **Dioscorea collettii** Hook.f. var. *collettii*

*Dioscorea septemloba* Griseb. = **Dioscorea sinuata** Vell.

*Dioscorea septemloba* var. *platyphylla* M.Mizush. ex T.Shimizu = **Dioscorea septemloba** Thunb.

*Dioscorea siamensis* R.Knuth = **Dioscorea glabra** Roxb.

*Dioscorea sieberi* Kunth = **Dioscorea polygonoides** Humb. & Bonpl. ex Willd.

*Dioscorea sikkimensis* Prain & Burkill = **Dioscorea prazeri** Prain & Burkill

*Dioscorea silvestris* Vell. = **Dioscorea amaranthoides** C.Presl

*Dioscorea similis* R.Knuth = **Dioscorea monadelpha** (Kunth) Griseb.

*Dioscorea sinuata* var. *bonariensis* Hauman = **Dioscorea sinuata** Vell.

*Dioscorea sinuata* var. *macrotepala* Uline = **Dioscorea sinuata** Vell.

*Dioscorea sinuata* var. *pauloensis* R.Knuth = **Dioscorea sinuata** Vell.

*Dioscorea sititoana* Honda & Jôtani = **Dioscorea septemloba** Thunb.

*Dioscorea soror* Prain & Burkill = **Dioscorea divaricata** Blanco

*Dioscorea soror* var. *glauca* Prain & Burkill = **Dioscorea divaricata** Blanco

*Dioscorea soror* var. *vera* Prain & Burkill = **Dioscorea divaricata** Blanco

*Dioscorea sororia* Kunth = **Dioscorea glandulosa** (Griseb.) Klotzsch ex Kunth var. *glandulosa*

*Dioscorea sparsiflora* Hemsl. = **Dioscorea remotiflora** Kunth

*Dioscorea spicata* (Vell.) Pedralli = **Dioscorea marginata** Griseb.

*Dioscorea spicata* var. *anamallayana* Prain & Burkill = **Dioscorea spicata** Roth

*Dioscorea spicata* var. *parvifolia* Prain & Burkill = **Dioscorea spicata** Roth

*Dioscorea spiculata* Blume = (Menispermaceae)

*Dioscorea spiculiflora* var. *chiapasana* Gómez Pompa = **Dioscorea gomez-pompae** O.Téllez

*Dioscorea spiculiflora* var. *fasciculocongesta* Sosa & B.G.Schub. = **Dioscorea fasciculocongesta** (Sosa & B.G.Schub.) O.Téllez

*Dioscorea spinosa* Roxb. ex Hook.f. = **Dioscorea esculenta** (Lour.) Burkill

*Dioscorea spinosa* Burm. = **Dioscorea pentaphylla** L.

*Dioscorea spiralis* Lem. = ?

*Dioscorea stellatopilosa* De Wild. = **Dioscorea schimperiana** Hochst. ex Kunth

*Dioscorea stellatopilosa* var. *cordata* De Wild. = **Dioscorea schimperiana** Hochst. ex Kunth

*Dioscorea stenophylla* var. *paucinervis* Uline ex R.Knuth = **Dioscorea stenophylla** Uline

*Dioscorea stolzii* R.Knuth = **Dioscorea cochleariapiculata** De Wild.

*Dioscorea stuhlmannii* Harms = **Dioscorea quartiniana** A.Rich.

*Dioscorea subcalva* var. *submollis* (R.Knuth) C.T.Ting & P.P.Ling = **Dioscorea subcalva** Prain & Burkill

*Dioscorea subfusca* R.Knuth = **Dioscorea kamoonensis** Kunth

*Dioscorea submollis* R.Knuth = **Dioscorea subcalva** Prain & Burkill

*Dioscorea sulcata* R.Knuth = **Dioscorea acanthogene** Rusby

*Dioscorea sulcata* R.Knuth = **Dioscorea glandulosa** (Griseb.) Klotzsch ex Kunth var. *glandulosa*

*Dioscorea sumbawensis* R.Knuth = [42 LSI]

*Dioscorea surinamensis* Miq. ex Knuth = **Dioscorea amazonum** Mart. ex Griseb. var. *amazonum*

*Dioscorea suthni* Buch.-Ham. = [40 IND]

*Dioscorea swinhoei* Rolfe = **Dioscorea polystachya** Turcz.

*Dioscorea sylvatica* var. *brevipes* (Burtt Davy) Burkill = **Dioscorea sylvatica** Eckl.

*Dioscorea sylvatica* subsp. *lydenbergensis* Blunden, Hardman & F.J.Hind = **Dioscorea sylvatica** Eckl.

*Dioscorea sylvatica* var. *multiflora* (Marloth) Burkill = **Dioscorea sylvatica** Eckl.

*Dioscorea sylvatica* var. *paniculata* (Dummer) Burkill = **Dioscorea sylvatica** Eckl.

*Dioscorea sylvatica* var. *rehmannii* (Baker) Burkill = **Dioscorea sylvatica** Eckl.

*Dioscorea sylvestris* De Wild. = **Dioscorea bulbifera** L.

*Dioscorea synandra* (Uline) Standl. = **Dioscorea gaumeri** R.Knuth

*Dioscorea tabascana* Matuda = **Dioscorea hondurensis** R.Knuth

*Dioscorea tafiensis* R.Knuth = **Dioscorea hieronymi** Uline ex R.Knuth

*Dioscorea tambillensis* R.Knuth = **Dioscorea moyobambensis** R.Knuth

*Dioscorea tamifolia* Salisb. = **Dioscorea bulbifera** L.

*Dioscorea tamifolia* Chodat & Hassl. = **Dioscorea pilcomayensis** Hauman

*Dioscorea tamoidea* var. *lindenii* Uline = **Dioscorea tamoidea** Griseb.

*Dioscorea tarokoensis* Hayata = **Dioscorea benthamii** Prain & Burkill

*Dioscorea tashiroi* Hayata = **Dioscorea collettii** Hook.f. var. *collettii*

*Dioscorea tenii* R.Knuth = **Dioscorea melanophyma** Prain & Burkill

*Dioscorea tenuiflora* Schltdl. = **Dioscorea bulbifera** L.

*Dioscorea tenuifolia* Uline ex R.Knuth = **Dioscorea warmingii** R.Knuth

*Dioscorea tepinapensis* Uline ex R.Knuth = **Dioscorea composita** Hemsl.

*Dioscorea tepinapensis* var. *aggregata* Uline ex R.Knuth = **Dioscorea composita** Hemsl.

*Dioscorea teretiuscula* Griseb. = **Dioscorea cinnamomifolia** Hook.

*Dioscorea testudinaria* R.Knuth = **Dioscorea elephantipes** (L'Hér.) Engl.

*Dioscorea therezopolensis* var. *latifolia* Uline ex R. Knuth = **Dioscorea therezopolensis** Uline ex R.Knuth

*Dioscorea thermarum* Phil. = **Dioscorea reticulata** Gay

*Dioscorea thinophila* Phil. = ***Dioscorea fastigiata*** Gay

*Dioscorea thonneri* De Wild. & T.Durand = ***Dioscorea preussii*** Pax subsp. ***preussii***

*Dioscorea tiliifolia* Kunth = ***Dioscorea esculenta*** (Lour.) Burkill

*Dioscorea tolucana* (Matuda) Caddick & Wilkin = ***Dioscorea multinervis*** Benth.

*Dioscorea toxicaria* Bojer = ***Dioscorea sansibarensis*** Pax

*Dioscorea triandria* Sessé & Moç. = ?

*Dioscorea triangularis* (Griseb.) R.Knuth = ***Dioscorea glandulosa*** (Griseb.) Klotzsch ex Kunth var. ***glandulosa***

*Dioscorea triangularis* Sessé & Moç. = ?

*Dioscorea trichoneura* Phil. = ***Dioscorea aristolochiifolia*** Poepp.

*Dioscorea trichopoda* Jum. & H.Perrier = ***Dioscorea soso*** var. ***trichopoda*** (Jum. & H.Perrier) Burkill & H.Perrier

*Dioscorea trifoliata* var. *amazonica* R.Knuth = ***Dioscorea trifoliata*** Kunth

*Dioscorea trifoliata* var. *galipanensis* (Kunth) Uline ex R.Knuth = ***Dioscorea trifoliata*** Kunth

*Dioscorea trilinguis* var. *edwallii* Uline ex R.Knuth = ***Dioscorea trilinguis*** Griseb.

*Dioscorea triloba* Lam. = ***Dioscorea trifida*** L.f.

*Dioscorea triloba* Willd. = ***Dioscorea trifida*** L.f.

*Dioscorea triloba* H.Karst. ex Kunth = ***Dioscorea trifoliata*** Kunth

*Dioscorea triphylla* Schimp. ex A.Rich. = ***Dioscorea dumetorum*** (Kunth) Pax

*Dioscorea triphylla* L. = ***Dioscorea pentaphylla*** L.

*Dioscorea triphylla* var. *abyssinica* R.Knuth = ***Dioscorea dumetorum*** (Kunth) Pax

*Dioscorea triphylla* var. *daemona* (Roxb.) Prain & Burkill = ***Dioscorea hispida*** Dennst.

*Dioscorea triphylla* var. *dumetorum* (Kunth) R.Knuth = ***Dioscorea dumetorum*** (Kunth) Pax

*Dioscorea triphylla* var. *mollissima* (Blume) Prain & Burkill = ***Dioscorea hispida*** Dennst.

*Dioscorea triphylla* var. *reticulata* (Hook.f.) Prain & Burkill = ***Dioscorea hispida*** Dennst.

*Dioscorea triphylla* var. *rotundata* R.Knuth = ***Dioscorea dumetorum*** (Kunth) Pax

*Dioscorea triphylla* var. *rotundata* R.Knuth = ***Dioscorea dumetorum*** (Kunth) Pax

*Dioscorea triphylla* var. *tomentosa* Rendle = ***Dioscorea dumetorum*** (Kunth) Pax

*Dioscorea truncata* R.H.Schomb. ex Prain = ***Dioscorea megacarpa*** Gleason

*Dioscorea truncata* Rusby = ***Dioscorea pseudorajanioides*** R.Knuth

*Dioscorea tuberosa* Vell. = ***Dioscorea cinnamomifolia*** Hook.

*Dioscorea tuberosa* Rojas Acosta = (Asparagaceae)

*Dioscorea tucumanensis* R.Knuth = ***Dioscorea stenopetala*** Hauman

*Dioscorea tuerckheimii* R.Knuth = ***Dioscorea mexicana*** Scheidw.

*Dioscorea tugui* Blanco = ***Dioscorea esculenta*** (Lour.) Burkill

*Dioscorea tuxtlensis* Sessé & Moç. = ?

*Dioscorea tweediei* R.Knuth = ***Dioscorea microbotrya*** Griseb.

*Dioscorea tysonii* Baker = ***Dioscorea retusa*** Mast.

*Dioscorea ulinei* var. *longipes* Matuda = ***Dioscorea ulinei*** Greenm. ex R.Knuth

*Dioscorea ulugurensis* R.Knuth = ***Dioscorea quartiniana*** A.Rich.

*Dioscorea undecimnervis* Vell. = ***Dioscorea glandulosa*** (Griseb.) Klotzsch ex Kunth var. ***glandulosa***

*Dioscorea undulata* R.Knuth = ***Dioscorea collettii*** var. ***hypoglauca*** (Palib.) S.J.Pei & C.T.Ting

*Dioscorea urceolata* f. *atropurpureoloba* Matuda = ***Dioscorea urceolata*** Uline

*Dioscorea urceolata* var. *reflexa* Greenm. ex R.Knuth = ***Dioscorea urceolata*** Uline

*Dioscorea vargasii* Standl. = ***Dioscorea incayensis*** R.Knuth

*Dioscorea variifolia* Kunze = ***Dioscorea sinuata*** Vell.

*Dioscorea velutina* Jum. & H.Perrier = ***Dioscorea ovinala*** Baker

*Dioscorea venosa* Uline ex R.Knuth = ***Dioscorea fodinarum*** Kunth

*Dioscorea venosa* var. *effusa* Uline = ***Dioscorea fodinarum*** Kunth

*Dioscorea venosa* var. *fodinarum* Uline = ***Dioscorea fodinarum*** Kunth

*Dioscorea verdickii* De Wild. = ***Dioscorea quartiniana*** A.Rich.

*Dioscorea vespertilio* Benth. = ***Dioscorea quartiniana*** A.Rich.

*Dioscorea villosa* var. *coreana* Prain & Burkill = ***Dioscorea coreana*** (Prain & Burkill) R.Knuth

*Dioscorea villosa* subsp. *floridana* (Bartlett) R.Knuth = ***Dioscorea floridana*** Bartlett

*Dioscorea villosa* var. *floridana* (Bartlett) H.E.Ahles = ***Dioscorea floridana*** Bartlett

*Dioscorea villosa* var. *glabra* J.Lloyd ex A.Gray = ***Dioscorea villosa*** L.

*Dioscorea villosa* subsp. *glabrifolia* (Bartlett) W.Stone = ***Dioscorea villosa*** L.

*Dioscorea villosa* subsp. *glabrifolia* (Bartlett) S.F.Blake = ***Dioscorea villosa*** L.

*Dioscorea villosa* f. *glabrifolia* (Bartlett) Fernald = ***Dioscorea villosa*** L.

*Dioscorea villosa* subsp. *glauca* (Muhl. ex L.C.Beck) R.Knuth = ***Dioscorea villosa*** L.

*Dioscorea villosa* subsp. *hirticaulis* (Bartlett) R.Knuth = ***Dioscorea villosa*** L.

*Dioscorea villosa* var. *hirticaulis* (Bartlett) H.E.Ahles = ***Dioscorea villosa*** L.

*Dioscorea villosa* var. *laeviuscula* Alph.Wood = ***Dioscorea villosa*** L.

*Dioscorea villosa* subsp. *paniculata* (Michx.) R.Knuth = ***Dioscorea villosa*** L.

*Dioscorea villosa* subsp. *quaternata* (Walter) R.Knuth = ***Dioscorea villosa*** L.

*Dioscorea violacea* Baudon = ***Dioscorea bulbifera*** L.

*Dioscorea violacea* Uline = ***Dioscorea dugesii*** C.B.Rob.

*Dioscorea violacea* R.Knuth = ***Dioscorea toldosensis*** R.Knuth

*Dioscorea vittata* W.Bull ex Baker = ***Dioscorea dodecaneura*** Vell.

*Dioscorea vulgaris* Miq. = ***Dioscorea alata*** L.

*Dioscorea wallichii* var. *christiei* Prain & Burkill = ***Dioscorea wallichii*** Hook.f.

*Dioscorea wallichii* var. *vera* Prain & Burkill = ***Dioscorea wallichii*** Hook.f.

*Dioscorea waltheri* Desf. = ***Dioscorea villosa*** L.

*Dioscorea welwitschii* Rendle = ***Dioscorea sansibarensis*** Pax

*Dioscorea wichurae* Uline ex R.Knuth = ***Dioscorea tokoro*** Makino ex Miyabe

*Dioscorea wilkesii* Uline ex R.Knuth = ***Dioscorea loheri*** Prain & Burkill

*Dioscorea yokusai* Prain & Burkill = ***Dioscorea tokoro*** Makino ex Miyabe

*Dioscorea yucatanensis* Uline = *Dioscorea matagalpensis* Uline

*Dioscorea zanoniae* Klotzsch ex Griseb. = *Dioscorea cinnamomifolia* Hook.

*Dioscorea zara* Baudon = *Dioscorea sagittifolia* var. *lecardii* (De Wild.) Nkounkou

*Dioscorea zemaroana* Koidz. = [38 NNS]

*Dioscorea zollingeriana* Kunth = *Dioscorea pyrifolia* Kunth var. *pyrifolia*

**Unplaced Names:**

*Dioscorea aculeata* L., Sp. Pl.: 1033 (1753), provisional synonym. = [40 IND] Perhaps identical with Dioscorea esculenta.

*Dioscorea angustifolia* Lam., Encycl. 3: 233 (1789). = [83 PER]

*Dioscorea boridiensis* R.Knuth, Repert. Spec. Nov. Regni Veg. 42: 163 (1937). = [43 NWG]

*Dioscorea clemensii* R.Knuth, Repert. Spec. Nov. Regni Veg. 40: 224 (1936). = [42 BOR]

*Dioscorea dallmannensis* Hatus., Bot. Mag. (Tokyo) 56: 425 (1942). = [43 NWG]

*Dioscorea dejantiana* Lem., Hort. Universel 5: 355 (1844). = ?

*Dioscorea ×hybrida* G.Nicholson, Ill. Dict. Gard. 4: 537 (1888). = *D. communis* × *D. polystachya*

*Dioscorea kweichowensis* R.Knuth, Repert. Spec. Nov. Regni Veg. 36: 125 (1934). = [36 CHC]

*Dioscorea owenii* Prain & Burkill, Bull. Misc. Inform. Kew 1925: 63 (1925). = [42 MLY]

*Dioscorea perforata* Bertero ex Spreng., Syst. Veg. 2: 152 (1825). = [81 LEE]

*Dioscorea spiralis* Lem., Hort. Universel 5: 355 (1844). = ?

*Dioscorea sumbawensis* R.Knuth, Repert. Spec. Nov. Regni Veg. 38: 119 (1935). = [42 LSI]

*Dioscorea suthni* Buch.-Ham. in V.H.Jackson (ed.), Account Distr. Purnea: 359 (1928). = [40 IND]

*Dioscorea triandria* Sessé & Moç., Pl. Nov. Hisp.: 172 (1890). = ?

*Dioscorea triangularis* Sessé & Moç., Fl. Mexic., ed. 2: 232 (1894). = ?

*Dioscorea tuxtlensis* Sessé & Moç., Fl. Mexic., ed. 2: 231 (1894). = ?

*Dioscorea zemaroana* Koidz., Pl. Nov. Amami-Ohsim.: 14 (1928). = [38 NNS]

## Elephantodon

*Elephantodon* Salisb. = *Dioscorea* Plum. ex L.

*Elephantodon eburnea* (Lour.) Salisb. = *Dioscorea alata* L.

## Epipetrum

*Epipetrum* Phil. = *Dioscorea* Plum. ex L.

*Epipetrum bilobum* Phil. = *Dioscorea biloba* (Phil.) Caddick & Wilkin

*Epipetrum humile* (Bertero ex Colla) Phil. = *Dioscorea humilis* Bertero ex Colla

*Epipetrum polyanthes* F.Phil. = *Dioscorea polyanthes* (F.Phil.) Caddick & Wilkin

## Halloschulzia

*Halloschulzia* Kuntze = *Stenomeris* Planch.

*Halloschulzia cumingiana* (Becc.) Kuntze = *Stenomeris dioscoreifolia* Planch.

*Halloschulzia dioscoreifolia* (Planch.) Kuntze = *Stenomeris dioscoreifolia* Planch.

## Hamatris

*Hamatris* Salisb. = *Dioscorea* Plum. ex L.

*Hamatris triphylla* (L.) Salisb. = *Dioscorea pentaphylla* L.

## Helmia

*Helmia* Kunth = *Dioscorea* Plum. ex L.

*Helmia adenocarpa* (Mart. ex Griseb.) Kunth = *Dioscorea ovata* Vell.

*Helmia anomala* (Griseb.) Kunth = *Dioscorea anomala* Griseb.

*Helmia aspera* (Humb. & Bonpl. ex Willd.) Kunth = *Dioscorea aspera* Humb. & Bonpl. ex Willd.

*Helmia brachycarpa* (Schltdl.) Kunth = *Dioscorea convolvulacea* Cham. & Schltdl. subsp. *convolvulacea*

*Helmia bulbifera* (L.) Kunth = *Dioscorea bulbifera* L.

*Helmia campestris* (Griseb.) Kunth = *Dioscorea campestris* Griseb.

*Helmia consanguinea* Kunth = *Dioscorea amazonum* Mart. ex Griseb. var. *amazonum*

*Helmia convolvulacea* (Cham. & Schltdl.) Kunth = *Dioscorea convolvulacea* Cham. & Schltdl.

*Helmia coriacea* (Humb. & Bonpl. ex Willd.) Kunth = *Dioscorea coriacea* Humb. & Bonpl. ex Willd.

*Helmia cuspidata* (Humb. & Bonpl. ex Willd.) Kunth = *Dioscorea cuspidata* Humb. & Bonpl. ex Willd.

*Helmia daemona* (Roxb.) Kunth = *Dioscorea hispida* Dennst.

*Helmia dregeana* Kunth = *Dioscorea dregeana* (Kunth) T.Durand & Schinz

*Helmia dumetorum* Kunth = *Dioscorea dumetorum* (Kunth) Pax

*Helmia ehrenbergiana* Kunth = *Dioscorea remotiflora* Kunth

*Helmia fracta* (Griseb.) Kunth = *Dioscorea furcata* Griseb.

*Helmia furcata* (Griseb.) Kunth = *Dioscorea furcata* Griseb.

*Helmia galipanensis* Kunth = *Dioscorea trifoliata* Kunth

*Helmia grisebachii* Kunth = *Dioscorea trisecta* Griseb.

*Helmia hirsuta* (Blume) Kunth = *Dioscorea hispida* Dennst.

*Helmia lacerdaei* (Griseb.) Kunth = *Dioscorea lacerdaei* Griseb.

*Helmia monodelpha* Kunth = *Dioscorea monadelpha* (Kunth) Griseb.

*Helmia moritziana* Kunth = *Dioscorea moritziana* (Kunth) R.Knuth

*Helmia multiflora* (Mart. ex Griseb.) Kunth = *Dioscorea multiflora* Mart. ex Griseb.

*Helmia pilosiuscula* (Bertero ex Spreng.) Kunth = *Dioscorea pilosiuscula* Bertero ex Spreng.

*Helmia psilostachya* Kunth = *Rajania psilostachya* (Kunth) Uline ex R.Knuth

*Helmia racemosa* Klotzsch = *Dioscorea racemosa* (Klotzsch) Uline

*Helmia scabra* (Humb. & Bonpl. ex Willd.) Kunth = *Dioscorea scabra* Humb. & Bonpl. ex Willd.

*Helmia schomburgkiana* Kunth = *Dioscorea pilosiuscula* Bertero ex Spreng.

*Helmia syringifolia* Kunth = *Dioscorea syringifolia* (Kunth) Kunth & R.H.Schomb. ex R.Knuth

*Helmia tomentosa* (J.König ex Spreng.) Kunth = *Dioscorea tomentosa* J.König ex Spreng.

*Helmia trifoliata* (Kunth) Kunth = *Dioscorea trifoliata* Kunth

## Higinbothamia

*Higinbothamia* Uline = *Dioscorea* Plum. ex L.
*Higinbothamia synandra* Uline = *Dioscorea gaumeri* R.Knuth

## Hyperocarpa

*Hyperocarpa* (Uline) G.M.Barroso, E.F.Guim. & Sucre = *Dioscorea* Plum. ex L.
*Hyperocarpa filiformis* G.M.Barroso, E.F.Guim. & Sucre = *Dioscorea grisebachii* Kunth

## Leontopetaloides

*Leontopetaloides* Boehm. = *Tacca* J.R.Forst. & G.Forst.

## Merione

*Merione* Salisb. = *Dioscorea* Plum. ex L.
*Merione peltata* (Burm.f.) Salisb. = (Menispermaceae)
*Merione villosa* (L.) Salisb. = *Dioscorea villosa* L.

## Nanarepenta

*Nanarepenta* Matuda = *Dioscorea* Plum. ex L.
*Nanarepenta guerrerensis* Matuda = *Dioscorea longirhiza* Caddick & Wilkin
*Nanarepenta juxtlahuacensis* O.Téllez & Dávila = *Dioscorea juxtlahuacensis* (O.Téllez & Dávila) Caddick & Wilkin
*Nanarepenta tolucana* Matuda = *Dioscorea multinervis* Benth.

## Oncorhiza

*Oncorhiza* Pers. = *Dioscorea* Plum. ex L.
*Oncorhiza esculentus* (Lour.) Pers. = *Dioscorea esculenta* (Lour.) Burkill

## Oncus

*Oncus* Lour. = *Dioscorea* Plum. ex L.
*Oncus esculentus* Lour. = *Dioscorea esculenta* (Lour.) Burkill

## Podianthus

*Podianthus* Schnizl. = *Trichopus* Gaertn.
*Podianthus arifolius* Schnizl. = *Trichopus zeylanicus* Gaertn.

## Polynome

*Polynome* Salisb. = *Dioscorea* Plum. ex L.
*Polynome alata* (L.) Salisb. = *Dioscorea alata* L.
*Polynome bulbifera* (L.) Salisb. = *Dioscorea bulbifera* L.

## Raja

*Raja* Burm. = *Rajania* L.

## Rajania

*Rajania* L., Sp. Pl.: 1032 (1753).
Caribbean. 81 BAH CUB DOM DUB HAI JAM LEE PUE WIN.
19 Species
The transfer of the species of *Rajania* to *Dioscorea* remains unpublished (see introduction). *Rajania* and its species are therefore presented as accepted names in this checklist.
*Raja* Burm. in C. Plumier, Pl. Amer. 148 (1758).

***Rajania angustifolia*** Sw., Prodr.: 59 (1788).
Cuba to Hispaniola. 81 CUB HAI. Cl.

***Rajania cephalocarpa*** Uline ex R.Knuth, Notizbl. Bot. Gart. Berlin-Dahlem 7: 221 (1917).
W. Cuba. 81 CUB.
*Rajania hermannii* R.Knuth, Notizbl. Bot. Gart. Berlin-Dahlem 7: 220 (1917).

***Rajania cordata*** L., Sp. Pl.: 1032 (1753). *Rajania cordata* var. *eucordata* R.Knuth in H.G.A.Engler (ed.), Pflanzenr., IV, 43: 334 (1924), nom. inval.
Caribbean. 81 CUB DOM JAM LEE PUE WIN. Cl. tuber geophyte.
*Rajania pleioneura* Griseb., Fl. Brit. W. I.: 588 (1864).
*Dioscorea scorpioidea* C.Wright, Anales Acad. Ci. Méd. Habana 8: 74 (1871).
*Rajania sintenisii* Uline in I.Urban, Symb. Antill. 3: 281 (1902).
*Dioscorea cyclophylla* Urb., Symb. Antill. 6: 4 (1909).
*Rajania cyclophylla* (Urb.) R.Knuth, Notizbl. Bot. Gart. Berlin-Dahlem 7: 218 (1917).
*Rajania cordata* var. *cymulifera* Uline, Notizbl. Bot. Gart. Berlin-Dahlem 7: 220 (1917).
*Rajania cordata* var. *microcarpa* Uline, Notizbl. Bot. Gart. Berlin-Dahlem 7: 219 (1917).
*Rajania cordata* var. *retusa* R.Knuth, Notizbl. Bot. Gart. Berlin-Dahlem 7: 219 (1917).
*Rajania cordata* var. *scorpioidea* R.Knuth, Notizbl. Bot. Gart. Berlin-Dahlem 7: 220 (1917).
*Rajania venosa* R.Knuth, Notizbl. Bot. Gart. Berlin-Dahlem 7: 219 (1917).

***Rajania ekmanii*** R.Knuth, Repert. Spec. Nov. Regni Veg. 21: 80 (1925).
Cuba. 81 CUB.

***Rajania hastata*** L., Sp. Pl.: 1032 (1753).
Hispaniola. 81 DOM HAI.
*Rajania hastata* var. *angusta* R.Knuth, Notizbl. Bot. Gart. Berlin-Dahlem 7: 222 (1917).
*Rajania hastata* var. *euhastata* R.Knuth, Notizbl. Bot. Gart. Berlin-Dahlem 7: 222 (1917).
*Rajania hastata* var. *incisa* R.Knuth, Notizbl. Bot. Gart. Berlin-Dahlem 7: 222 (1917).
*Rajania hastata* var. *latior* R.Knuth, Notizbl. Bot. Gart. Berlin-Dahlem 7: 222 (1917).
*Rajania hastata* var. *triloba* R.Knuth, Notizbl. Bot. Gart. Berlin-Dahlem 7: 222 (1917).

***Rajania microphylla*** Kunth, Enum. Pl. 5: 451 (1850).
Bahamas, Cuba. 81 BAH CUB.
*Rajania bahamensis* R.Knuth, Notizbl. Bot. Gart. Berlin-Dahlem 7: 220 (1917).
*Rajania urbaniana* R.Knuth, Notizbl. Bot. Gart. Berlin-Dahlem 7: 220 (1917).
*Rajania prestoniensis* R.Knuth in H.G.A.Engler (ed.), Pflanzenr., IV, 43: 337 (1924).

***Rajania minutiflora*** Uline ex R.Knuth, Notizbl. Bot. Gart. Berlin-Dahlem 7: 221 (1917).
Haiti. 81 HAI.

***Rajania nipensis*** R.A.Howard, J. Arnold Arbor. 28: 117 (1947).
E. Cuba. 81 CUB.

***Rajania ovata*** Sw., Prodr.: 59 (1788).
E. Cuba to Hispaniola. 81 CUB DOM HAI.
*Rajania cordata* var. *ehrenbergii* Uline, Notizbl. Bot. Gart. Berlin-Dahlem 7: 219 (1917).
*Rajania ovata* var. *ehrenbergii* R.Knuth, Notizbl. Bot. Gart. Berlin-Dahlem 7: 219 (1917).
*Dioscorea haitiensis* R.Knuth, Ark. Bot. 20A(5): 6 (1926).

*Rajania pilifera* Urb., Ark. Bot. 17(7): 17 (1922).
Haiti. 81 HAI.

*Rajania porulosa* R.Knuth, Notizbl. Bot. Gart. Berlin-
Dahlem 7: 221 (1917).
EC. & E. Cuba. 81 CUB.

*Rajania psilostachya* (Kunth) Uline ex R.Knuth,
Notizbl. Bot. Gart. Berlin-Dahlem 7: 218 (1917).
Cuba. 81 CUB.
*Helmia psilostachya* Kunth, Enum. Pl. 5: 429 (1850).

*Rajania quinquefolia* L., Sp. Pl.: 1032 (1753).
C. & E. Cuba to Hispaniola. 81 DOM DUB HAI.
*Rajania mucronata* Willd., Sp. Pl. 4: 787 (1806).
*Rajania angustifolia* Griseb. ex Kunth, Enum. Pl. 5:
446 (1850).
*Rajania cubensis* Kunth, Enum. Pl. 6: 446 (1850).
*Dioscorea lanceolata* C.Wright ex Griseb., Cat. Pl.
Cub.: 251 (1866).

*Rajania spiculiflora* Uline ex R.Knuth, Notizbl. Bot.
Gart. Berlin-Dahlem 7: 219 (1917).
Haiti. 81 HAI.
*Rajania marginata* R.Knuth, Repert. Spec. Nov. Regni
Veg. 38: 121 (1935).

*Rajania tenella* R.A.Howard, J. Arnold Arbor. 28: 119
(1947).
Cuba. 81 CUB.

*Rajania tenuiflora* R.Knuth, Notizbl. Bot. Gart. Berlin-
Dahlem 7: 219 (1917).
E. Cuba. 81 CUB. Cl.
*Rajania baracoensis* R.Knuth in H.G.A.Engler (ed.),
Pflanzenr., IV, 43: 333 (1924).

*Rajania theresensis* Uline ex R.Knuth, Notizbl. Bot.
Gart. Berlin-Dahlem 7: 221 (1917).
Cuba (I. Teresa). 81 CUB.

*Rajania wilsoniana* C.V.Morton, Proc. Biol. Soc. Wash.
46: 85 (1933).
C. Cuba. 81 CUB. Cl.

*Rajania wrightii* Uline ex R.Knuth, Notizbl. Bot. Gart.
Berlin-Dahlem 7: 221 (1917).
Cuba to Hispaniola. 81 CUB DOM. Cl.
*Rajania hastata* Griseb. in C.F.P.von Martius & auct.
suc. (eds.), Fl. Bras. 3(1): 47 (1842), nom. illeg.

**Synonyms:**
*Rajania angustifolia* Griseb. ex Kunth = *Rajania*
*quinquefolia* L.
*Rajania bahamensis* R.Knuth = *Rajania microphylla*
Kunth
*Rajania baracoensis* R.Knuth = *Rajania tenuiflora*
R.Knuth
*Rajania brasiliensis* Griseb. = *Dioscorea cinnamomifolia*
Hook.
*Rajania cirrhata* Billb. ex Beurl. = [80 COS]
*Rajania cordata* var. *cymulifera* Uline = *Rajania*
*cordata* L.
*Rajania cordata* var. *ehrenbergii* Uline = *Rajania ovata*
Sw.
*Rajania cordata* var. *eucordata* R.Knuth = *Rajania*
*cordata* L.
*Rajania cordata* var. *microcarpa* Uline = *Rajania*
*cordata* L.
*Rajania cordata* var. *retusa* R.Knuth = *Rajania cordata*
L.
*Rajania cordata* var. *scorpioidea* R.Knuth = *Rajania*
*cordata* L.
*Rajania cubensis* Kunth = *Rajania quinquefolia* L.

*Rajania cyclophylla* (Urb.) R.Knuth = *Rajania cordata*
L.
*Rajania flexuosa* Poir. = *Dioscorea bridgesii* Griseb. ex
Kunth
*Rajania hastata* Griseb. = *Rajania wrightii* Uline ex
R.Knuth
*Rajania hastata* var. *angusta* R.Knuth = *Rajania hastata*
L.
*Rajania hastata* var. *euhastata* R.Knuth = *Rajania*
*hastata* L.
*Rajania hastata* var. *incisa* R.Knuth = *Rajania hastata*
L.
*Rajania hastata* var. *latior* R.Knuth = *Rajania hastata* L.
*Rajania hastata* var. *triloba* R.Knuth = *Rajania hastata*
L.
*Rajania hermannii* R.Knuth = *Rajania cephalocarpa*
Uline ex R.Knuth
*Rajania herradurensis* R.Knuth = *Dioscorea wrightii*
Uline ex R.Knuth
*Rajania hexaphylla* Thunb. = (Lardizabalaceae)
*Rajania linearis* R.A.Howard = *Dioscorea nipensis*
R.A.Howard
*Rajania lobata* Poir. = [83 PER]
*Rajania marginata* R.Knuth = *Rajania spiculiflora* Uline
ex R.Knuth
*Rajania mucronata* Willd. = *Rajania quinquefolia* L.
*Rajania ovata* var. *ehrenbergii* R.Knuth = *Rajania ovata*
Sw.
*Rajania pleioneura* Griseb. = *Rajania cordata* L.
*Rajania prestoniensis* R.Knuth = *Rajania microphylla*
Kunth
*Rajania quinata* Thunb. ex Houtt. = (Lardizabalaceae)
*Rajania quinquenervia* Raf. = [81 CUB]
*Rajania sintenisii* Uline = *Rajania cordata* L.
*Rajania urbaniana* R.Knuth = *Rajania microphylla*
Kunth
*Rajania venosa* R.Knuth = *Rajania cordata* L.

**Unplaced Names:**
*Rajania cirrhata* Billb. ex Beurl., Kongl. Vetensk. Acad.
Handl. 1854: 111 (1856). = [80 COS]
*Rajania lobata* Poir. in J.B.A.M.de Lamarck, Encycl. 6:
58 (1804). = [83 PER]
*Rajania quinquenervia* Raf., Autik. Bot.: 125 (1840). =
[81 CUB]

## Rhizemys

*Rhizemys* Raf. = *Dioscorea* Plum. ex L.
*Rhizemys elephantipes* (L'Hér.) Raf. = *Dioscorea*
*elephantipes* (L'Hér.) Engl.
*Rhizemys montana* (Burch.) Raf. = *Dioscorea*
*elephantipes* (L'Hér.) Engl.

## Ricophora

*Ricophora* Mill. = *Dioscorea* Plum. ex L.

## Schizocapsa

*Schizocapsa* Hance = *Tacca* J.R.Forst. & G.Forst.
*Schizocapsa breviscapa* (Ostenf.) H.Limpr. = *Tacca*
*chantrieri* André
*Schizocapsa guangxiensis* P.P.Ling & C.T.Ting = *Tacca*
*plantaginea* (Hance) Drenth
*Schizocapsa itagakii* Yamam. = *Tacca chantrieri* André
*Schizocapsa plantaginea* Hance = *Tacca plantaginea*
(Hance) Drenth

## Sismondaea

*Sismondaea* Delponte = ***Dioscorea*** Plum. ex L.
*Sismondaea dioscoreoides* Delponte = ***Dioscorea piperifolia*** Humb. & Bonpl. ex Willd.

## Steireya

*Steireya* Raf. = ***Trichopus*** Gaertn.
*Steireya angustifolia* (Lindl.) Raf. = ***Trichopus zeylanicus*** Gaertn.
*Steireya cordata* (Lindl.) Raf. = ***Trichopus zeylanicus*** Gaertn.
*Steireya media* Raf. = ***Trichopus zeylanicus*** Gaertn.

## Stenomeris

***Stenomeris*** Planch., Ann. Sci. Nat., Bot., III, 18: 319 (1852).
W. & C. Malesia. 42 BOR MLY PHI SUM.
2 Species
*Halloschulzia* Kuntze, Revis. Gen. Pl. 2: 705 (1891).

***Stenomeris borneensis*** Oliv., Hooker's Icon. Pl. 24: t. 2328 (1894).
W. Malesia to Philippines (Mindanao). 42 BOR MLY PHI SUM. Cl. rhizome geophyte.
*Stenomeris mindanaensis* R.Knuth in H.G.A.Engler (ed.), Pflanzenr., IV, 43: 346 (1924).

***Stenomeris dioscoreifolia*** Planch., Ann. Sci. Nat., Bot., III, 18: 320 (1852). *Halloschulzia dioscoreifolia* (Planch.) Kuntze, Revis. Gen. Pl. 2: 705 (1891).
Philippines. 42 PHI. Cl. rhizome geophyte.
*Stenomeris cumingiana* Becc., Nuovo Giorn. Bot. Ital. 2: 8 (1870). *Halloschulzia cumingiana* (Becc.) Kuntze, Revis. Gen. Pl. 2: 705 (1891).
*Stenomeris wallisii* Taub., Bot. Jahrb. Syst. 15(38): 2 (1893).

**Synonyms:**
*Stenomeris cumingiana* Becc. = ***Stenomeris dioscoreifolia*** Planch.
*Stenomeris mindanaensis* R.Knuth = ***Stenomeris borneensis*** Oliv.
*Stenomeris wallisii* Taub. = ***Stenomeris dioscoreifolia*** Planch.

## Strophis

*Strophis* Salisb. = ***Dioscorea*** Plum. ex L.
*Strophis cirrhosa* (Lour.) Salisb. = ***Dioscorea cirrhosa*** Lour.

## Tacca

***Tacca*** J.R.Forst. & G.Forst., Char. Gen. Pl.: 35 (1775), nom. cons.
Trop. & Subtrop. 22 BEN BKN GAM GHA GNB GUI IVO MLI NGA NGR SEN SIE TOG 23 BUR CAF CMN CON GAB ZAI 24 CHA ETH SUD 25 KEN TAN UGA 26 ANG MLW MOZ ZAM ZIM 29 COM MAU MDG SEY 36 CHC CHH CHS CHT (38) tai 40 ASS BAN EHM IND LDV MDV PAK SRL 41 AND CBD LAO MYA THA VIE 42 BOR JAW LSI MLY MOL PHI SUL SUM 43 BIS NWG SOL 50 NTA QLD WAU 60 FIJ GIL nue NWC SAM TOK TON VAN WAL 61 COO MRQ PIT SCI TUA TUB 62 CRL MRN MRS (63) haw 82 GUY SUR? VEN 83 CLM PER 84 BZE BZN.
13 Species
*Leontopetaloides* Boehm., Defin. Gen. Pl.: 512 (1760).
*Ataccia* C.Presl, Reliq. Haenk. 1: 149 (1828).

*Chaitaea* Sol. ex Seem., Fl. Vit.: 102 (1866).
*Schizocapsa* Hance, J. Bot. 19: 292 (1881).

***Tacca ankaranensis*** Bard.-Vauc., Adansonia, III, 19: 154 (1997).
Madagascar. 29 MDG.

***Tacca bibracteata*** Drenth, Blumea 20: 395 (1972 publ. 1973).
Borneo (Sarawak). 42 BOR.

***Tacca celebica*** Koord., Meded. Lands Plantentuin 19: 641 (1898).
N. Sulawesi. 42 SUL. Rhizome geophyte.
*Tacca minahassae* Koord., Meded. Lands Plantentuin 19: 641 (1898).

***Tacca chantrieri*** André, Rev. Hort. 73: 541 (1901).
Assam to S. China and Pen. Malaysia. 36 CHC CHH CHS CHT 40 ASS BAN 41 CBD LAO MYA THA VIE 42 MLY. Rhizome geophyte.
*Tacca macrantha* H.Limpr., Beitr. Kenttn. Tacca: 45 (1902).
*Tacca lancifolia* var. *breviscapa* Ostenf., Bot. Tidsskr. 26: 165 (1904). *Schizocapsa breviscapa* (Ostenf.) H.Limpr. in H.G.A.Engler (ed.), Pflanzenr., IV, 42: 11 (1928).
*Tacca minor* Ridl., Mat. Fl. Malay. Penins. 2: 78 (1907).
*Tacca vespertilio* Ridl., Mat. Fl. Malay. Penins. 2: 77 (1907).
*Clerodendrum esquirolii* H.Lév., Repert. Spec. Nov. Regni Veg. 11: 298 (1912), nom. illeg. *Tacca esquirolii* Rehder, J. Arnold Arbor. 17: 64 (1936).
*Tacca garrettii* Craib, Bull. Misc. Inform. Kew 1912: 106 (1912).
*Tacca paxiana* H.Limpr. in H.G.A.Engler (ed.), Pflanzenr., IV, 42: 16 (1928).
*Tacca roxburghii* H.Limpr. in H.G.A.Engler (ed.), Pflanzenr., IV, 42: 18 (1928).
*Tacca wilsonii* H.Limpr., Repert. Spec. Nov. Regni Veg. 38: 218 (1935).
*Schizocapsa itagakii* Yamam., Contr. Fl. Hainan 1: 32 (1942).

***Tacca ebeltajae*** Drenth, Blumea 20: 401 (1972 publ. 1973).
E. New Guinea to Solomon Is. 43 NWG SOL. Tuber geophyte.

***Tacca integrifolia*** Ker Gawl., Bot. Mag. 35: t. 1448 (1812). *Ataccia integrifolia* (Ker Gawl.) C.Presl, Reliq. Haenk. 1: 149 (1828).
Bhutan to W. Malesia. 36 CHT 40 ASS BAN EHM IND PAK 41 CBD LAO MYA THA VIE 42 BOR JAW MLY SUM. Hemicr. or rhizome geophyte.
*Tacca cristata* Jack, Malayan Misc. 1(5): 23 (1820). *Ataccia cristata* (Jack) Kunth, Enum. Pl. 5: 466 (1850).
*Tacca aspera* Roxb., Fl. Ind. 2: 169 (1824). *Ataccia aspera* (Roxb.) Kunth, Enum. Pl. 5: 464 (1850).
*Tacca laevis* Roxb., Fl. Ind. 2: 171 (1824). *Ataccia laevis* (Roxb.) Kunth, Enum. Pl. 5: 466 (1850).
*Tacca rafflesiana* Jack ex Wall., Numer. List: 5172 (1831), nom. inval.
*Tacca lancifolia* Zoll. & Moritzi, Syst. Verz.: 91 (1846). *Ataccia lancifolia* (Zoll. & Moritzi) Kunth, Enum. Pl. 5: 465 (1850).
*Tacca borneensis* Ridl., J. Straits Branch Roy. Asiat. Soc. 49: 45 (1908).
*Tacca sumatrana* H.Limpr. in H.G.A.Engler (ed.), Pflanzenr., IV, 42: 18 (1928).

*Tacca choudhuriana* Deb, Indian Forester 90: 241 (1964).

**Tacca leontopetaloides** (L.) Kuntze, Revis. Gen. Pl. 2: 704 (1891).
Trop. Old World to Pacific. 22 BEN BKN GAM GHA GNB GUI IVO MLI NGA NGR SEN SIE TOG 23 BUR CAF CMN CON GAB ZAI 24 CHA ETH SUD 25 KEN TAN UGA 26 ANG MLW MOZ ZAM ZIM 29 COM MAU MDG SEY (38) tai 40 IND LDV MDV SRL 41 AND CBD LAO MYA THA VIE 42 BOR JAW LSI MLY MOL PHI SUL SUM 43 BIS NWG SOL 50 NTA QLD WAU 60 FIJ GIL nue NWC SAM TOK TON VAN WAL 61 COO MRQ PIT SCI TUA TUB 62 CRL MRN MRS (63) haw. Tuber geophyte.
*Tacca pinnatifida* J.R.Forst. & G.Forst., Char. Gen. Pl.: 35 (1775).
*Tacca pinnatifolia* Gaertn., Fruct. Sem. Pl. 1: 43 (1788).
*Tacca involucrata* Schumach. & Thonn. in C.F. Schumacher, Beskr. Guin. Pl.: 177 (1827). *Tacca pinnatifida* subsp. *involucrata* (Schumach. & Thonn.) H.Limpr., Beitr. Kenttn. Tacca: 55 (1902).
*Tacca dubia* Schult. & Schult.f. in J.J.Roemer & J.A.Schultes, Syst. Veg. 7: 167 (1829).
*Tacca phallifera* Schult. & Schult.f. in J.J.Roemer & J.A.Schultes, Syst. Veg. 7: 167 (1829).
*Tacca gaogao* Blanco, Fl. Filip.: 262 (1837).
*Tacca madagascariensis* Bojer, Hortus Maurit.: 350 (1837).
*Tacca maculata* Zipp. ex Span., Linnaea 15: 480 (1841), nom. inval.
*Tacca quanzensis* Welw., Apont.: 591 (1859).
*Chaitaea tacca* Sol. ex Seem., Fl. Vit.: 102 (1866).
*Tacca artocarpifolia* Seem., Fl. Vit.: 101 (1866).
*Tacca brownii* Seem., Fl. Vit.: 100 (1866).
*Tacca oceanica* Seem., J. Bot. 4: 261 (1866).
*Tacca viridis* Hemsl., Hooker's Icon. Pl. 26: t. 2515 (1897).
*Tacca abyssinica* Hochst. ex Baker in D.Oliver & auct. suc. (eds.), Fl. Trop. Afr. 7: 413 (1898).
*Tacca pinnatifida* var. *acutifolia* H.Limpr., Beitr. Kenttn. Tacca: 55 (1902). *Tacca involucrata* var. *acutifolia* (H.Limpr.) H.Limpr. in H.G.A.Engler (ed.), Pflanzenr., IV, 42: 29 (1928).
*Tacca pinnatifida* subsp. *interrupta* Warb. ex H.Limpr., Beitr. Kenttn. Tacca: 56 (1902).
*Tacca pinnatifida* subsp. *madagascariensis* H.Limpr., Beitr. Kenttn. Tacca: 55 (1902). *Tacca madagascariensis* (H.Limpr.) H.Limpr. in H.G.A.Engler (ed.), Pflanzenr., IV, 42: 29 (1928).
*Tacca umbrarum* Jum. & H.Perrier, Ann. Inst. Bot.-Géol. Colon. Marseille, II, 8: 386 (1910).
*Tacca pinnatifida* var. *paeoniifolia* Domin, Biblioth. Bot. 85: 533 (1915).
*Tacca pinnatifida* var. *permagna* Domin, Biblioth. Bot. 85: 532 (1915).
*Tacca hawaiiensis* H.Limpr. in H.G.A.Engler (ed.), Pflanzenr., IV, 42: 30 (1928).

**Tacca maculata** Seem., Fl. Vit.: 103 (1866). *Tacca pinnatifida* var. *aconitifolia* F.Muell. ex Benth., Fl. Austral. 6: 459 (1873). *Tacca pinnatifida* subsp. *maculata* (Seem.) H.Limpr., Beitr. Kenttn. Tacca: 56 (1902). *Tacca pinnatifida* var. *maculata* (Seem.) Domin, Biblioth. Bot. 20: 534 (1915), nom. superfl.
NE. Western Australia to NW. Northern Territory, SW. Pacific. 50 NTA WAU 60 FIJ SAM. Tuber geophyte.
*Tacca samoensis* Reinecke, Bot. Jahrb. Syst. 25: 595 (1898).

**Tacca palmata** Blume, Enum. Pl. Javae 1: 83 (1827).
Indo-China to Malesia (incl. Misool I.). 41 CBD THA VIE 42 BOR JAW LSI MLY MOL PHI SUL SUM 43 NWG. Tuber geophyte.
*Tacca montana* Schult. & Schult.f. in J.J.Roemer & J.A.Schultes, Syst. Veg. 7: 168 (1829).
*Tacca vesicaria* Blanco, Fl. Filip.: 261 (1837).
*Tacca rumphii* Schauer, Nov. Actorum Acad. Caes. Leop.-Carol. Nat. Cur. 19(Suppl. 1): 442 (1843).
*Tacca elmeri* K.Krause, Leafl. Philipp. Bot. 6: 2283 (1914).
*Tacca angustilobata* Merr., Philipp. J. Sci. 29: 356 (1926).
*Tacca fatsiifolia* Warb. ex H.Limpr. in H.G.A.Engler (ed.), Pflanzenr., IV, 42: 23 (1928).

**Tacca palmatifida** Baker, J. Linn. Soc., Bot. 15: 100 (1876).
Sulawesi. 42 SUL. Rhizome geophyte.
*Tacca flabellata* J.J.Sm., Bull. Jard. Bot. Buitenzorg, III, 6: 79 (1924).
*Tacca breviloba* Warb. ex H.Limpr. in H.G.A.Engler (ed.), Pflanzenr., IV, 42: 22 (1928).

**Tacca parkeri** Seem., Fl. Vit.: 102 (1866).
S. Trop. America. 82 GUY SUR? VEN 83 CLM PER 84 BZE BZN. Hemicr.
*Tacca lanceolata* Seem., Fl. Vit.: 102 (1866). *Tacca parkeri* var. *lanceolata* (Seem.) H.Limpr. in H.G.A.Engler (ed.), Pflanzenr., IV, 42: 21 (1928).
*Tacca sprucei* Benth. in G.Bentham & J.D.Hooker, Gen. Pl. 3: 741 (1883).
*Tacca parkeri* f. *paraensis* H.Limpr. in H.G.A.Engler (ed.), Pflanzenr., IV, 42: 21 (1928).
*Tacca ulei* H.Limpr. in H.G.A.Engler (ed.), Pflanzenr., IV, 42: 22 (1928).

**Tacca plantaginea** (Hance) Drenth, Blumea 20: 391 (1972 publ. 1973).
S. China to Indo-China. 36 CHC CHH CHS 41 LAO THA VIE. Hemicr. or rhizome geophyte.
*Schizocapsa plantaginea* Hance, J. Bot. 19: 292 (1881).
*Schizocapsa guangxiensis* P.P.Ling & C.T.Ting, Acta Phytotax. Sin. 20: 202 (1982).

**Tacca subflabellata** P.P.Ling & C.T.Ting, Acta Phytotax. Sin. 20: 202 (1982).
China (SE. Yunnan). 36 CHC. Hemicr. or rhizome geophyte.

### Synonyms:

*Tacca abyssinica* Hochst. ex Baker = **Tacca leontopetaloides** (L.) Kuntze
*Tacca angustilobata* Merr. = **Tacca palmata** Blume
*Tacca artocarpifolia* Seem. = **Tacca leontopetaloides** (L.) Kuntze
*Tacca aspera* Roxb. = **Tacca integrifolia** Ker Gawl.
*Tacca borneensis* Ridl. = **Tacca integrifolia** Ker Gawl.
*Tacca breviloba* Warb. ex H.Limpr. = **Tacca palmatifida** Baker
*Tacca brownii* Seem. = **Tacca leontopetaloides** (L.) Kuntze
*Tacca choudhuriana* Deb = **Tacca integrifolia** Ker Gawl.
*Tacca cristata* Jack = **Tacca integrifolia** Ker Gawl.
*Tacca dubia* Schult. & Schult.f. = **Tacca leontopetaloides** (L.) Kuntze
*Tacca elmeri* K.Krause = **Tacca palmata** Blume
*Tacca esquirolii* Rehder = **Tacca chantrieri** André
*Tacca fatsiifolia* Warb. ex H.Limpr. = **Tacca palmata** Blume
*Tacca flabellata* J.J.Sm. = **Tacca palmatifida** Baker
*Tacca gaogao* Blanco = **Tacca leontopetaloides** (L.) Kuntze
*Tacca garrettii* Craib = **Tacca chantrieri** André

*Tacca hawaiiensis* H.Limpr. = ***Tacca leontopetaloides*** (L.) Kuntze

*Tacca involucrata* Schumach. & Thonn. = ***Tacca leontopetaloides*** (L.) Kuntze

*Tacca involucrata* var. *acutifolia* (H.Limpr.) H.Limpr. = ***Tacca leontopetaloides*** (L.) Kuntze

*Tacca laevis* Roxb. = ***Tacca integrifolia*** Ker Gawl.

*Tacca lanceolata* Seem. = ***Tacca parkeri*** Seem.

*Tacca lancifolia* Zoll. & Moritzi = ***Tacca integrifolia*** Ker Gawl.

*Tacca lancifolia* var. *breviscapa* Ostenf. = ***Tacca chantrieri*** André

*Tacca macrantha* H.Limpr. = ***Tacca chantrieri*** André

*Tacca maculata* Zipp. ex Span. = ***Tacca leontopetaloides*** (L.) Kuntze

*Tacca madagascariensis* Bojer = ***Tacca leontopetaloides*** (L.) Kuntze

*Tacca madagascariensis* (H.Limpr.) H.Limpr. = ***Tacca leontopetaloides*** (L.) Kuntze

*Tacca minahassae* Koord. = ***Tacca celebica*** Koord.

*Tacca minor* Ridl. = ***Tacca chantrieri*** André

*Tacca montana* Schult. & Schult.f. = ***Tacca palmata*** Blume

*Tacca oceanica* Seem. = ***Tacca leontopetaloides*** (L.) Kuntze

*Tacca parkeri* var. *lanceolata* (Seem.) H.Limpr. = ***Tacca parkeri*** Seem.

*Tacca parkeri* f. *paraensis* H.Limpr. = ***Tacca parkeri*** Seem.

*Tacca paxiana* H.Limpr. = ***Tacca chantrieri*** André

*Tacca phallifera* Schult. & Schult.f. = ***Tacca leontopetaloides*** (L.) Kuntze

*Tacca pinnatifida* J.R.Forst. & G.Forst. = ***Tacca leontopetaloides*** (L.) Kuntze

*Tacca pinnatifida* var. *aconitifolia* F.Muell. ex Benth. = ***Tacca maculata*** Seem.

*Tacca pinnatifida* var. *acutifolia* H.Limpr. = ***Tacca leontopetaloides*** (L.) Kuntze

*Tacca pinnatifida* subsp. *interrupta* Warb. ex H.Limpr. = ***Tacca leontopetaloides*** (L.) Kuntze

*Tacca pinnatifida* subsp. *involucrata* (Schumach. & Thonn.) H.Limpr. = ***Tacca leontopetaloides*** (L.) Kuntze

*Tacca pinnatifida* subsp. *maculata* (Seem.) H.Limpr. = ***Tacca maculata*** Seem.

*Tacca pinnatifida* var. *maculata* (Seem.) Domin = ***Tacca maculata*** Seem.

*Tacca pinnatifida* subsp. *madagascariensis* H.Limpr. = ***Tacca leontopetaloides*** (L.) Kuntze

*Tacca pinnatifida* var. *paeoniifolia* Domin = ***Tacca leontopetaloides*** (L.) Kuntze

*Tacca pinnatifida* var. *permagna* Domin = ***Tacca leontopetaloides*** (L.) Kuntze

*Tacca pinnatifolia* Gaertn. = ***Tacca leontopetaloides*** (L.) Kuntze

*Tacca quanzensis* Welw. = ***Tacca leontopetaloides*** (L.) Kuntze

*Tacca rafflesiana* Jack ex Wall. = ***Tacca integrifolia*** Ker Gawl.

*Tacca roxburghii* H.Limpr. = ***Tacca chantrieri*** André

*Tacca rumphii* Schauer = ***Tacca palmata*** Blume

*Tacca samoensis* Reinecke = ***Tacca maculata*** Seem.

*Tacca sprucei* Benth. = ***Tacca parkeri*** Seem.

*Tacca sumatrana* H.Limpr. = ***Tacca integrifolia*** Ker Gawl.

*Tacca ulei* H.Limpr. = ***Tacca parkeri*** Seem.

*Tacca umbrarum* Jum. & H.Perrier = ***Tacca leontopetaloides*** (L.) Kuntze

*Tacca vesicaria* Blanco = ***Tacca palmata*** Blume

*Tacca vespertilio* Ridl. = ***Tacca chantrieri*** André

*Tacca viridis* Hemsl. = ***Tacca leontopetaloides*** (L.) Kuntze

*Tacca wilsonii* H.Limpr. = ***Tacca chantrieri*** André

## Tamnus

*Tamnus* Mill. = ***Dioscorea*** Plum. ex L.

## Tamus

*Tamus* L. = ***Dioscorea*** Plum. ex L.

*Tamus baccifera* St.-Lag. = ***Dioscorea communis*** (L.) Caddick & Wilkin

*Tamus canariensis* Willd. ex Kunth = ***Dioscorea communis*** (L.) Caddick & Wilkin

*Tamus cirrhosa* Hausskn. ex Bornm. = ***Dioscorea communis*** (L.) Caddick & Wilkin

*Tamus communis* L. = ***Dioscorea communis*** (L.) Caddick & Wilkin

*Tamus communis* subsp. *cretica* (L.) Nyman = ***Dioscorea communis*** (L.) Caddick & Wilkin

*Tamus communis* var. *cretica* (L.) Boiss. = ***Dioscorea communis*** (L.) Caddick & Wilkin

*Tamus communis* var. *subtriloba* Guss. = ***Dioscorea communis*** (L.) Caddick & Wilkin

*Tamus communis* f. *subtriloba* (Guss.) O.Bolòs & Vigo = ***Dioscorea communis*** (L.) Caddick & Wilkin

*Tamus communis* var. *triloba* Simonk. = ***Dioscorea communis*** (L.) Caddick & Wilkin

*Tamus cordifolia* Stokes = ***Dioscorea communis*** (L.) Caddick & Wilkin

*Tamus cretica* L. = ***Dioscorea communis*** (L.) Caddick & Wilkin

*Tamus edulis* Lowe = ***Dioscorea communis*** (L.) Caddick & Wilkin

*Tamus elephantipes* L'Hér. = ***Dioscorea elephantipes*** (L'Hér.) Engl.

*Tamus nepalensis* Jacquem. ex Prain & Burkill = ***Dioscorea deltoidea*** Wall. ex Griseb.

*Tamus norsa* Lowe = ***Dioscorea communis*** (L.) Caddick & Wilkin

*Tamus orientalis* J.Thiébaut = ***Dioscorea orientalis*** (J.Thiébaut) Caddick & Wilkin

*Tamus parviflora* Kunth = ***Dioscorea communis*** (L.) Caddick & Wilkin

*Tamus racemosa* Gouan = ***Dioscorea communis*** (L.) Caddick & Wilkin

*Tamus sylvestris* Kunth = ***Dioscorea sylvatica*** Eckl.

## Testudinaria

*Testudinaria* Salisb. ex Burch. = ***Dioscorea*** Plum. ex L.

*Testudinaria cocolmeca* Procop. = ***Dioscorea mexicana*** Scheidw.

*Testudinaria elephantipes* (L'Hér.) Burch. = ***Dioscorea elephantipes*** (L'Hér.) Engl.

*Testudinaria elephantipes* f. *montana* (Burch.) G.D.Rowley = ***Dioscorea elephantipes*** (L'Hér.) Engl.

*Testudinaria macrostachya* (Benth.) G.D.Rowley = ***Dioscorea mexicana*** Scheidw.

*Testudinaria montana* Burch. = ***Dioscorea elephantipes*** (L'Hér.) Engl.

*Testudinaria montana* var. *paniculata* Kuntze = ***Dioscorea sylvatica*** Eckl.

*Testudinaria multiflora* Marloth = ***Dioscorea sylvatica*** Eckl.

*Testudinaria paniculata* Dummer = ***Dioscorea sylvatica*** Eckl.

*Testudinaria rehmannii* (Baker) G.D.Rowley = ***Dioscorea sylvatica*** Eckl.

*Testudinaria sylvatica* (Eckl.) Kunth = ***Dioscorea sylvatica*** Eckl.

*Testudinaria sylvatica* var. *brevipes* (Burtt Davy) G.D.Rowley = ***Dioscorea sylvatica*** Eckl.

*Testudinaria sylvatica* var. *lydenbergensis* (Blunden, Hardman & F.J.Hind) G.D.Rowley = ***Dioscorea sylvatica*** Eckl.

*Testudinaria sylvatica* var. *multiflora* (Marloth) G.D. Rowley = ***Dioscorea sylvatica*** Eckl.

*Testudinaria sylvatica* var. *paniculata* (Dummer)
G.D.Rowley = **Dioscorea sylvatica** Eckl.
*Testudinaria sylvatica* var. *rehmannii* (Baker)
G.D.Rowley = **Dioscorea sylvatica** Eckl.

## Trichopodium

*Trichopodium* Lindl. = **Trichopus** Gaertn.
*Trichopodium angustifolium* Lindl. = **Trichopus zeylanicus** Gaertn.
*Trichopodium cordatum* Lindl. = **Trichopus zeylanicus** Gaertn.
*Trichopodium intermedium* Lindl. = **Trichopus zeylanicus** Gaertn.
*Trichopodium travancoricum* Bedd. = **Trichopus zeylanicus** Gaertn.
*Trichopodium zeylanicum* (Gaertn.) Thwaites = **Trichopus zeylanicus** Gaertn.

## Trichopus

**Trichopus** Gaertn., Fruct. Sem. Pl. 1: 44 (1788).
Madagascar, SW. India to Pen. Malaysia. 29 MDG 40 IND SRL 41 THA 42 MLY.
2 Species
*Trichopodium* Lindl., Edwards's Bot. Reg. 18: t. 1543 (1832).
*Steireya* Raf., Fl. Tellur. 4: 100 (1838).
*Podianthus* Schnizl., Bot. Zeitung (Berlin) 1: 739 (1843).
*Avetra* H.Perrier, Bull. Soc. Bot. France 71: 25 (1924).

**Trichopus sempervirens** (H.Perrier) Caddick & Wilkin, Taxon 51: 113 (2002).
E. Madagascar. 29 MDG. Cl. rhizome geophyte.
*\*Avetra sempervirens* H.Perrier, Bull. Soc. Bot. France 71: 27 (1924).

**Trichopus zeylanicus** Gaertn., Fruct. Sem. Pl. 1: 44 (1788). *Trichopodium zeylanicum* (Gaertn.) Thwaites, Enum. Pl. Zeyl.: 291 (1861).
SW. India to Pen. Malaysia. 40 IND SRL 41 THA 42 MLY. Rhizome geophyte.
*Trichopodium angustifolium* Lindl., Edwards's Bot. Reg. 18: t. 1543 (1832). *Steireya angustifolia* (Lindl.) Raf., Fl. Tellur. 4: 100 (1838). *Trichopus zeylanicus* var. *angustifolius* (Lindl.) R.Knuth in H.G.A.Engler (ed.), Pflanzenr., IV, 43: 348 (1924). *Trichopus zeylanicus* subsp. *angustifolius* (Lindl.) Sivar., Pushp. & P.K.R.Kumar, Kew Bull. 45: 357 (1990).

*Trichopodium cordatum* Lindl., Edwards's Bot. Reg. 18: t. 1543 (1832). *Steireya cordata* (Lindl.) Raf., Fl. Tellur. 4: 100 (1838). *Trichopus zeylanicus* var. *cordatus* (Lindl.) R.Knuth in H.G.A.Engler (ed.), Pflanzenr., IV, 43: 348 (1924).
*Trichopodium intermedium* Lindl., Edwards's Bot. Reg. 19: t. 1543 (1832). *Trichopus zeylanicus* var. *intermedius* (Lindl.) R.Knuth in H.G.A.Engler (ed.), Pflanzenr., IV, 43: 348 (1924).
*Steireya media* Raf., Fl. Tellur. 4: 100 (1838).
*Podianthus arifolius* Schnizl., Bot. Zeitung (Berlin) 1: 739 (1843).
*Trichopodium travancoricum* Bedd., Icon. Pl. Ind. Or.: 68 (1874). *Trichopus zeylanicus* subsp. *travancoricus* (Bedd.) Burkill, in Fl. Males. 4: 299 (1951).
*Trichopus malayanus* Ridl., Fl. Malay Penins. 4: 312 (1924).

**Synonyms:**
*Trichopus malayanus* Ridl. = **Trichopus zeylanicus** Gaertn.
*Trichopus piperifolius* Wall. ex Griff. = (Aristolochiaceae)
*Trichopus zeylanicus* var. *angustifolius* (Lindl.) R.Knuth = **Trichopus zeylanicus** Gaertn.
*Trichopus zeylanicus* subsp. *angustifolius* (Lindl.) Sivar., Pushp. & P.K.R.Kumar = **Trichopus zeylanicus** Gaertn.
*Trichopus zeylanicus* var. *cordatus* (Lindl.) R.Knuth = **Trichopus zeylanicus** Gaertn.
*Trichopus zeylanicus* subsp. *travancoricus* (Bedd.) Burkill = **Trichopus zeylanicus** Gaertn.

## Ubium

*Ubium* J.F.Gmel. = **Dioscorea** Plum. ex L.
*Ubium quadrifarium* J.F.Gmel. = **Dioscorea pentaphylla** L.
*Ubium scandens* J.St.-Hil. = **Dioscorea pentaphylla** L.

# Nartheciaceae

## Abama

*Abama* Adans., Fam. Pl. 2: 47 (1763). = *Narthecium* Huds.

*Abama americana* (Ker Gawl.) Morong, Mem. Torrey Bot. Club 5: 109 (1894). = *Narthecium americanum* Ker Gawl.

*Abama asiatica* (Maxim.) Makino, J. Jap. Bot. 7: 4 (1931). = *Narthecium asiaticum* Maxim.

*Abama californica* (Baker) A.Heller, Cat. N. Amer. Pl.: 3 (1898). = *Narthecium californicum* Baker

*Abama montana* Small, Torreya 24: 86 (1924). = *Narthecium americanum* Ker Gawl.

*Abama occidentalis* (A.Gray) A.Heller, Muhlenbergia 1: 47 (1904). = *Narthecium californicum* Baker

*Abama ossifraga* (L.) DC. in J.B.A.M.de Lamarck & A.P.de Candolle, Fl. Franç. 3: 171 (1805). = *Narthecium ossifragum* (L.) Huds.

## Aletris

*Aletris* L., Sp. Pl.: 319 (1753).
Arabian Pen., China to Japan and W. Malesia, E. Canada to E. U.S.A., Bahamas. 36 CHC CHN CHS CHT 38 JAP KOR NNS TAI 40 ASS EHM NEP WHM 41 MYA 42 BOR MLY? PHI SUM 72 ONT 74 ILL OKL WIS 75 CNT INI MAS MIC NWH NWJ NWY PEN RHO WVA 77 TEX 78 ALA ARK DEL FLA GEO KTY LOU MRY MSI NCA SCA TEN VRG WDC 81 BAH.
23 Species

*Aletris alpestris* Diels, Bot. Jahrb. Syst. 36(82): 20 (1905). C. China. 36 CHC CHN. Hemicr.

*Aletris aurea* Walter, Fl. Carol.: 121 (1788).
SE. U.S.A. to Texas. 74 OKL 77 TEX 78 ALA ARK FLA GEO LOU MRY MSI NCA SCA VRG. Hemicr.

*Aletris bracteata* Northr., Mem. Torrey Bot. Club 12: 27 (1902).
S. Florida, Bahamas. 78 FLA 81 BAH. Hemicr. or rhizome geophyte.

*Aletris capitata* F.T.Wang & Tang, in Fl. Reipubl. Popul. Sin. 15: 254 (1978).
China (C. Sichuan). 36 CHC. Hemicr.

*Aletris cinerascens* F.T.Wang & Tang, in Fl. Reipubl. Popul. Sin. 15: 254 (1978).
China (WC. Yunnan, Guangxi). 36 CHC CHS. Hemicr. or rhizome geophyte.

*Aletris farinosa* L., Sp. Pl.: 319 (1753).
SE. Canada to C. & E. U.S.A. 72 ONT 74 ILL OKL WIS 75 CNT INI MAI MAS MIC NWH NWJ NWY PEN RHO WVA 77 TEX 78 ALA ARK DEL FLA GEO KTY LOU MRY MSI NCA SCA TEN VRG WDC. Hemicr. or rhizome geophyte.

*Aletris foliata* (Maxim.) Makino & Nemoto, Fl. Japan, ed. 2: 1534 (1931).
Korea, Japan. 38 JAP KOR. Hemicr.
*Metanarthecium foliatum* Maxim.

*Aletris foliolosa* Stapf, Trans. Linn. Soc. London, Bot. 4: 240 (1894).
N. Sumatera, N. Borneo, Philippines (Mindoro). 42 BOR PHI SUM. Hemicr.

*Aletris foliosa* (Maxim.) Bureau & Franch., J. Bot. (Morot) 5: 156 (1891).
Japan (Hokkaido, N. & C. Honshu, Shikoku). 38 JAP. Hemicr.

*Aletris glabra* Bureau & Franch., J. Bot. (Morot) 5: 156 (1891).
Himalaya to China, Taiwan. 36 CHC CHN CHS CHT 38 KOR TAI 40 EHM NEP. Hemicr.

*Aletris glandulifera* Bureau & Franch., J. Bot. (Morot) 5: 156 (1891).
C. China. 36 CHC CHN. Hemicr.

*Aletris gracilis* Rendle, J. Bot. 44: 41 (1906).
E. Nepal to China (NW. Yunnan). 36 CHC CHT 40 EHM NEP 41 MYA. Hemicr.

*Aletris laxiflora* Bureau & Franch., J. Bot. (Morot) 5: 155 (1891).
E. Tibet to China (Sichuan, Guizhou). 36 CHC CHT. Hemicr.

*Aletris lutea* Small, Bull. New York Bot. Gard. 1: 278 (1899).
SE. U.S.A. 78 ALA FLA GEO LOU MSI. Hemicr.

*Aletris megalantha* F.T.Wang & Tang, Acta Phytotax. Sin. 1: 119 (1951).
China (W. Yunnan). 36 CHC. Hemicr.

*Aletris nana* S.C.Chen, Acta Phytotax. Sin. 19: 503 (1981).
Nepal, S. Tibet, China (NW. Yunnan). 36 CHC CHT 40 NEP. Hemicr.

*Aletris obovata* Nash, Torreya 3: 102 (1903).
SE. U.S.A. 78 ALA FLA GEO MSI SCA. Hemicr.

*Aletris pauciflora* (Klotzsch) Hand.-Mazz., Symb. Sin. 7: 1220 (1936).
Himalaya to SC. China. 36 CHC CHT 40 ASS EHM NEP WHM 41 MYA. Hemicr. or rhizome geophyte.
*Stachyopogon pauciflorus* Klotzsch

*Aletris pauciflora* var. **khasiana** (Hook.f.) F.T.Wang & Tang, in Fl. Reipubl. Popul. Sin. 15: 172 (1978).
Assam to SC. China. 36 CHC CHT 40 ASS 41 MYA. Hemicr.
*Aletris khasiana* Hook.f.

*Aletris pauciflora* var. **pauciflora**.
Himalaya to SC. China. 36 CHC CHT 40 EHM NEP WHM 41 MYA. Hemicr.

*Aletris pedicellata* F.T.Wang & Tang, Bull. Fan Mem. Inst. Biol. 1: 109 (1943).
China (Sichuan). 36 CHC. Hemicr.

*Aletris scopulorum* Dunn, J. Linn. Soc., Bot. 38: 370 (1908).
SE. China, Japan (Shikoku). 36 CHS 38 JAP. Hemicr.

*Aletris spicata* (Thunb.) Franch., J. Bot. (Morot) 10: 199
(1896).
China to SC. & C. Japan and Philippines (N. Luzon).
36 CHC CHN CHS 38 JAP KOR NNS TAI 42
MLY? PHI. Hemicr.
*Hypoxis spicata* Thunb.

*Aletris stenoloba* Franch., J. Bot. (Morot) 10: 203
(1896).
C. & S. China. 36 CHC CHN CHS. Hemicr.

*Aletris × tottenii* E.T.Br., Rhodora 63: 306 (1961). *A.
lutea × A. obovata*.
SE. U.S.A. 78 GEO. Hemicr.

*Aletris yaanica* G.H.Yang, Acta Phytotax. Sin. 25: 237
(1987).
China (C. Sichuan). 36 CHC. Hemicr.

**Synonyms:**
*Aletris alba* Michx., Fl. Bor.-Amer. 1: 189 (1803). =
**Aletris farinosa** L.
*Aletris alpestris* var. *occidentalis* H.Hara, J. Jap. Bot. 47:
276 (1972). = **Aletris nana** S.C.Chen
*Aletris biondiana* Diels, Bot. Jahrb. Syst. 36(82): 19
(1905). = **Aletris glandulifera** Bureau & Franch.
*Aletris brachyphylla* (Merr.) F.T.Wang & Tang, Bull. Fan
Mem. Inst. Biol. 7: 82 (1936). = **Aletris foliolosa**
Stapf
*Aletris dickinsii* Franch., Bull. Soc. Philom. Paris, VII,
10: 102 (1886). = **Aletris spicata** (Thunb.) Franch.
*Aletris dielsii* F.T.Wang & Tang, Bull. Fan Mem. Inst.
Biol. 7: 83 (1936). = **Aletris alpestris** Diels
*Aletris elata* F.T.Wang & Tang, Bull. Fan Mem. Inst.
Biol. 1: 108 (1943). = **Aletris laxiflora** Bureau &
Franch.
*Aletris farinosa* (Thunb.) Thunb., Fl. Jap.: 136 (1784),
nom. illeg. = **Aletris spicata** (Thunb.) Franch.
*Aletris fauriei* H.Lév. & Vaniot, Repert. Spec. Nov. Regni
Veg. 5: 283 (1908). = **Aletris foliata** (Maxim.) Makino
& Nemoto
*Aletris foliata* var. *glabra* (Bureau & Franch.) Yamam., J.
Soc. Trop. Agric. 10: 121 (1938). = **Aletris glabra**
Bureau & Franch.
*Aletris foliosa* var. *sikkimensis* (Hook.f.) Franch., J. Bot.
(Morot) 10: 198 (1896). = **Aletris glabra** Bureau &
Franch.
*Aletris formosana* (Hayata) Makino & Nemoto, Fl.
Japan, ed. 2: 1534 (1931). = **Aletris glabra** Bureau &
Franch.
*Aletris gracilipes* F.T.Wang & Tang, Bull. Fan Mem. Inst.
Biol. 1: 107 (1943). = **Aletris laxiflora** Bureau &
Franch.
*Aletris japonica* Lamb., Trans. Linn. Soc. London 10: 407
(1811), nom. illeg. = **Aletris spicata** (Thunb.) Franch.
*Aletris japonica* Thunb., Nova Acta Regiae Soc. Sci.
Upsal. 3: 208 (1780). = (Asparagaceae)
*Aletris khasiana* Hook.f., Fl. Brit. India 6: 265 (1892). =
**Aletris pauciflora** var. *khasiana* (Hook.f.) F.T.Wang
& Tang
*Aletris lactiflora* Franch., J. Bot. (Morot) 10: 200 (1896).
= **Aletris glandulifera** Bureau & Franch.
*Aletris lanuginosa* Bureau & Franch., J. Bot. (Morot) 5:
155 (1891). = **Aletris pauciflora** var. *khasiana*
(Hook.f.) F.T.Wang & Tang
*Aletris linguiformis* Burm.f., Fl. Indica, Prodr. Fl. Cap.:
10 (1768). = [27+]
*Aletris longibracteata* T.L.Xu, Acta Bot. Yunnan. 17: 21
(1995). = **Aletris stenoloba** Franch.
*Aletris lucida* Raf., Autik. Bot.: 136 (1840). = **Aletris
farinosa** L.

*Aletris lutea* f. *albiflora* E.T.Browne, Rhodora 63: 305
(1961). = **Aletris lutea** Small
*Aletris luteoviridis* (Maxim.) Franch., J. Bot. (Morot) 10:
201 (1896). = **Metanarthecium luteoviride** Maxim.
*Aletris mairei* H.Lév., Bull. Acad. Int. Géogr. Bot. 24: 37
(1915). = **Aletris pauciflora** var. *pauciflora*
*Aletris makiyataroi* Naruh., Acta Phytotax. Geobot. 25:
131 (1973). = **Aletris scopulorum** Dunn
*Aletris nepalensis* Hook.f., Fl. Brit. India 7: 264 (1892),
nom. illeg. = **Aletris pauciflora** (Klotzsch) Hand.-
Mazz.
*Aletris revoluta* Franch., J. Bot. (Morot) 10: 202 (1896).
= **Aletris laxiflora** Bureau & Franch.
*Aletris rigida* Stapf, Trans. Linn. Soc. London, Bot. 4:
241 (1894). = **Aletris foliolosa** Stapf
*Aletris sikkimensis* Hook.f., Fl. Brit. India 6: 265 (1892).
= **Aletris glabra** Bureau & Franch.
*Aletris spicata* var. *fargesii* Franch., J. Bot. (Morot) 10:
200 (1896). = **Aletris stenoloba** Franch.
*Aletris spicata* var. *microantha* Satake, J. Jap. Bot. 17:
725 (1941). = **Aletris spicata** (Thunb.) Franch.
*Aletris stelliflora* Hand.-Mazz., Symb. Sin. 7: 1219
(1936). = **Aletris gracilis** Rendle
*Aletris sumatrana* Masam., Bull. Soc. Bot. France 84: 18
(1937). = **Aletris foliolosa** Stapf
*Aletris tavelii* H.Lév., Cat. Pl. Yunnan: 130 (1916). =
**Aletris glabra** Bureau & Franch.

**Unplaced Names:**
*Aletris linguiformis* Burm.f., Fl. Indica, Prodr. Fl. Cap.:
10 (1768). = [27+]

## Lophiola

**Lophiola** Ker Gawl., Bot. Mag. 39: t. 1596 (1813).
E. Canada to E. U.S.A. 72 NSC 75 NWJ 78 ALA DEL
FLA GEO LOU MSI NCA.
1 Species

**Lophiola aurea** Ker Gawl., Bot. Mag. 39: t. 1596 (1813).
Nova Scotia to E. U.S.A. 72 NSC 75 NWJ 78 ALA DEL
FLA GEO LOU MSI NCA. Rhizome geophyte.

**Synonyms:**
*Lophiola americana* (Pursh) A.Wood, Class-book Bot.:
697 (1861). = **Lophiola aurea** Ker Gawl.
*Lophiola breviflora* Gand., Bull. Soc. Bot. France 66:
290 (1919 publ. 1920). = **Lophiola aurea** Ker Gawl.
*Lophiola floridana* Gand., Bull. Soc. Bot. France 66: 290
(1919 publ. 1920). = **Lophiola aurea** Ker Gawl.
*Lophiola septentrionalis* Fernald, Rhodora 23: 243
(1922). = **Lophiola aurea** Ker Gawl.
*Lophiola tomentosa* (Raf.) Britton, Sterns & Poggenb.,
Prelim. Cat.: 53 (1888). = **Lophiola aurea** Ker Gawl.

## Meta-aletris

*Meta-aletris* Masam., Trans. Nat. Hist. Soc. Taiwan 28:
46 (1938). = **Aletris** L.
*Meta-aletris rigida* (Stapf) Masam., Trans. Nat. Hist.
Soc. Taiwan 28: 46 (1938). = **Aletris foliolosa** Stapf
*Meta-aletris sumatrana* (Masam.) Masam., Trans. Nat.
Hist. Soc. Taiwan 28: 46 (1938). = **Aletris foliolosa**
Stapf

## Metanarthecium

**Metanarthecium** Maxim., Bull. Acad. Imp. Sci. Saint-
Pétersbourg 11: 438 (1867).
Kuril Is. to Japan, Korea. 31 KUR 38 JAP KOR.
1 Species

*Metanarthecium luteoviride* Maxim., Bull. Acad. Imp. Sci. Saint-Pétersbourg 11: 438 (1867).
Kuril Is. to Japan, Korea. 31 KUR 38 JAP KOR. Hemicr. or rhizome geophyte.

*Metanarthecium luteoviride* var. *luteoviride*.
Kuril Is. to Japan, Korea. 31 KUR 38 JAP KOR. Hemicr. or rhizome geophyte.

*Metanarthecium luteoviride* var. *nutans* Masam. in ?.
Japan (Yakushima). Rhizome geophyte.

*Synonyms*:
*Metanarthecium brachyphyllum* (Merr.) Masam., Bull. Soc. Bot. France 84: 18 (1937). = *Aletris foliolosa* Stapf
*Metanarthecium foliatum* Maxim. in E.R.von Trautvetter & al., Dec. Pl. Nov.: 10 (1882). = *Aletris foliata* (Maxim.) Makino & Nemoto
*Metanarthecium formosanum* Hayata, Icon. Pl. Formosan. 9: 142 (1920). = *Aletris glabra* Bureau & Franch.
*Metanarthecium yakumontanum* Masam., Trans. Nat. Hist. Soc. Taiwan 28: 115 (1938). = *Metanarthecium luteoviride* var. *nutans* Masam.

## Narthecium

*Narthecium* Huds., Fl. Angl.: 127 (1762), nom. cons.
Europe to Caucasus, Japan, W. & E. U.S.A. 10 DEN FOR GRB IRE NOR SWE 11 BGM CZE† GER NET 12 COR FRA POR SPA 13 ALB GRC YUG 33 TCS 34 TUR 38 JAP 73 ORE 75 NWJ 76 CAL 78 DEL MRY NCA SCA.
7 Species

*Narthecium americanum* Ker Gawl., Bot. Mag. 37: t. 1505 (1812).
New Jersey to South Carolina. 75 NWJ 78 DEL MRY NCA SCA. Hemicr. or rhizome geophyte.

*Narthecium asiaticum* Maxim., Bull. Acad. Imp. Sci. Saint-Pétersbourg 11: 438 (1867).
N. & C. Japan. 38 JAP. Rhizome geophyte.

*Narthecium balansae* Briq., Annuaire Conserv. Jard. Bot. Genève 5: 77 (1901).
NE. Turkey to Transcaucasus. 33 TCS 34 TUR. Rhizome geophyte.

*Narthecium californicum* Baker, J. Linn. Soc., Bot. 15: 351 (1876).
SW. Oregon to C. California. 73 ORE 76 CAL. Hemicr. or rhizome geophyte.

*Narthecium ossifragum* (L.) Huds., Fl. Angl.: 128 (1762).
SE. Sweden to N. Portugal. 10 DEN FOR GRB IRE NOR SWE 11 BGM CZE† GER NET 12 FRA POR SPA. Rhizome geophyte.
*Anthericum ossifragum* L.

*Narthecium reverchonii* Celak., Oesterr. Bot. Z. 37: 154 (1887).
Corse. 12 COR. Rhizome geophyte.

*Narthecium scardicum* Košanin, Oesterr. Bot. Z. 63: 141 (1913).
NE. Albania to NW. Greece. 13 ALB GRC YUG. Rhizome geophyte.

*Synonyms*:
*Narthecium anthericoides* Hoppe ex Mert. & W.D.J.Koch in J.C.Röhling, Deutschl. Fl., ed. 3, 2: 559 (1826). = *Narthecium ossifragum* (L.) Huds.

*Narthecium caucasicum* Miscz., Fl. Cauc. Crit. 2(4): 83 (1912). = *Narthecium balansae* Briq.
*Narthecium montanum* (Small) Grey, Hardy Bulbs 2: 446 (1938). = *Narthecium americanum* Ker Gawl.
*Narthecium occidentale* (A.Gray) Grey, Hardy Bulbs 2: 446 (1938). = *Narthecium californicum* Baker
*Narthecium ossifragum* var. *americanum* (Ker Gawl.) A.Gray, Manual, ed. 5: 536 (1867). = *Narthecium americanum* Ker Gawl.
*Narthecium ossifragum* var. *occidentale* A.Gray, Proc. Amer. Acad. Arts 7: 391 (1868). = *Narthecium californicum* Baker
*Narthecium palustre* Bubani, Fl. Pyren. 4: 169 (1902). = *Narthecium ossifragum* (L.) Huds.

## Nietneria

*Nietneria* Klotzsch ex Benth. in G.Bentham & J.D.Hooker, Gen. Pl. 3: 825 (1883).
SE. Venezuela to W. Guyana and N. Brazil. 82 GUY VEN 84 BZN.
2 Species

*Nietneria corymbosa* Klotzsch & M.R.Schomb. ex B.D.Jacks., Index Kew. 2: 314 (1894).
S. Venezuela. 82 VEN. Hemicr. or rhizome geophyte.

*Nietneria paniculata* Steyerm., Fieldiana, Bot. 28(1): 153 (1951).
SE. Venezuela to W. Guyana and N. Brazil (Serra Aracá). 82 GUY VEN 84 BZN. Rhizome geophyte.

## Stachyopogon

*Stachyopogon* Klotzsch, Bot. Ergebn. Reise Waldemar: 49 (1862). = *Aletris* L.
*Stachyopogon pauciflorus* Klotzsch, Bot. Ergebn. Reise Waldemar: 49 (1862). = *Aletris pauciflora* (Klotzsch) Hand.-Mazz.
*Stachyopogon spicatus* Klotzsch, Bot. Ergebn. Reise Waldemar: 49 (1862). = *Aletris pauciflora* var. *pauciflora*